Science and Technology

The HighScope Preschool Curriculum

Science and Technology

Ann S. Epstein, PhD

HIGHSCOPE PRESS®

Ypsilanti, Michigan

Published by
HighScope® Press

A division of the
HighScope Educational Research Foundation
600 North River Street
Ypsilanti, Michigan 48198-2898
734.485.2000, FAX 734.485.0704

Orders: 800.40.PRESS; Fax: 800.442.4FAX; www.highscope.org
E-mail: *press@highscope.org*

Copyright © 2012 by HighScope Educational Research Foundation. All rights reserved. Except as permitted under the Copyright Act of 1976, no part of this book may be reproduced or distributed in any form or by any means, electronic or mechanical, including photocopy, recording, or any information storage-and-retrieval system, without either the prior written permission from the publisher, or authorization through payment of the appropriate per-copy fee to the Copyright Clearance Center, Inc., 222 Rosewood Drive, Danvers, MA 01923, 978.750.8400, fax 978.646.8600, or on the web at www.copyright.com. The name "HighScope" and its corporate logos are registered trademarks and service marks of the HighScope Foundation.

Editor: Joanne Tangorra
Cover design, text design, production: Judy Seling, Seling Design LLC
Photography:
Bob Foran — Front cover, 1, 5, 6, 15, 17, 20 (KDI 45), 21 (KDI 50), 31, 34, 36, 39, 53, 54, 58, 75, 80, back cover
Gregory Fox — 9, 11 (top), 16, 20–21 (KDI 46, KDI 47, KDI 48, KDI 52), 25, 30, 44, 48, 57, 59, 61, 62, 64, 67, 70, 76, 82, 89, 90, 92, 96
Keansburg Public Schools, New Jersey — 28, 72
HighScope Staff — All other photos

Library of Congress Cataloging-in-Publication Data
Epstein, Ann S.
  Science and technology / Ann S. Epstein, PhD.
      pages cm. -- (The HighScope preschool curriculum)
  Includes bibliographical references.
  ISBN 978-1-57379-657-6 (soft cover : alk. paper) 1. Science--Study and teaching (Preschool) 2. Science--Curricula. I. Title.
  LB1140.5.S35E67 2012
  372.35'044--dc23
                                        2012003779

Printed in the United States of America
10 9 8 7 6 5 4 3 2 1

Contents

Acknowledgments vii

Chapter 1. The Importance of Science and Technology 1
What Is Science and Technology? 3
Components of Science and Technology 2
About This Book 10

Chapter 2: General Teaching Strategies for Science and Technology 15
General Teaching Strategies 16
    Introduce children to the steps in the scientific method 16
    Encourage reflection 17
    Create opportunities for surprise and discrepancy 18
    Encourage documentation 19
    Encourage collaborative investigation and problem solving 19
Key Developmental Indicators 22

Chapter 3: KDI 45. Observing 25
How Observing Develops 26
Teaching Strategies That Support Observing 27
    Provide a sensory-rich environment 27
    Establish a safe environment for children to observe with all their senses 29
    Provide the vocabulary to help children label, understand, and use their observations 30
Ideas for Scaffolding KDI 45. Observing 32

Chapter 4. KDI 46. Classifying 33
How Classifying Develops 35
Teaching Strategies That Support Classifying 37
    Encourage children to collect and sort things 37
    Call attention to "same" and "different" 38
    Use "no" and "not" language 40
Ideas for Scaffolding KDI 46. Classifying 42

Chapter 5. KDI 47. Experimenting 43
How Experimenting Develops 45
Teaching Strategies That Support Experimenting 48
    Ask and answer *"What if…?" "Why…?"* and *"How…?"* questions 48
    Encourage children to gradually replace trial-and-error exploration with systematic experimentation 49
    Provide materials and experiences for investigating how things change with time 50
Ideas for Scaffolding KDI 47. Experimenting 52

Chapter 6. KDI 48. Predicting 53
How Predicting Develops 55
Teaching Strategies That Support Predicting 56
    Help children reflect on the similarities between their past and present experiences 56
    Encourage children to say what they think will happen 56
    Encourage children to verify their predictions 58
Ideas for Scaffolding KDI 48. Predicting 60

Chapter 7. KDI 49. Drawing Conclusions 61

How Drawing Conclusions Develops 63

Teaching Strategies That Support Drawing Conclusions 65

    Provide materials and experiences that work in similar but not identical ways 65

    Encourage children to reflect on the processes and outcomes they observe 66

Ideas for Scaffolding KDI 49. Drawing Conclusions 68

Chapter 8. KDI 50. Communicating Ideas 69

How Communicating Ideas Develops 71

Teaching Strategies That Support Communicating Ideas 73

    Use scientific language as you talk with children about their actions, observations, and discoveries 73

    Provide opportunities for children to symbolically represent their scientific experiences 74

Ideas for Scaffolding KDI 50. Communicating Ideas 78

Chapter 9. KDI 51. Natural and Physical World 79

How Knowledge About the Natural and Physical World Develops 81

Teaching Strategies That Support Learning About the Natural and Physical World 83

    Provide materials and experiences for children to gather knowledge about the natural and physical world 83

    Encourage children to make connections to explain how the world looks and functions 84

Ideas for Scaffolding KDI 51. Natural and Physical World 87

Chapter 10. KDI 52. Tools and Technology 89

How Use of Tools and Technology Develops 91

    Experimenting with tools 91

    Using computers and other technology 91

Teaching Strategies That Support Use of Tools and Technology 93

    Provide a variety of tools in all areas of the classroom 93

    Help children consider how and why to use tools in various ways 94

    Choose and mediate children's use of appropriate technology 96

Ideas for Scaffolding KDI 52. Tools and Technology 98

References 101

Acknowledgments

Many people contributed their knowledge and skills to the publication of *Science and Technology*. I want to thank the early childhood and other staff members who collaborated on creating the key developmental indicators (KDIs) in this content area: Beth Marshall, Sue Gainsley, Shannon Lockhart, Polly Neill, Kay Rush, Julie Hoelscher, and Emily Thompson. Among this group of colleagues, those who devoted special attention to reviewing the manuscript for this book were Polly Neill and Sue Gainsley. Mary Hohmann, whose expertise informs many other HighScope Curriculum books, also provided detailed feedback.

The developmental scaffolding charts in this volume — describing what children might do and say and how adults can support and gently extend their learning at different developmental levels — are invaluable contributions to the curriculum. I am grateful to Beth Marshall and Sue Gainsley for the extraordinary working relationship we forged in creating these charts. By bringing our unique experiences to this challenging process, we integrated knowledge about child development and effective classroom practices from the perspectives of research, teaching, training, and policy.

Thanks are also due to Nancy Brickman, who directed the editing and production of the book. I extend particular appreciation to Joanne Tangorra, who edited the volume, and Katie Bruckner, who assisted with all aspects of the publication process. I also want to acknowledge the following individuals for contributing to the book's visual appeal and reader friendliness: photographers Bob Foran and Gregory Fox and graphic artists Judy Seling (book designer) and Kazuko Sacks (book production).

Finally, I extend sincerest thanks to all the teachers, trainers, children, and families whose participation in HighScope and other early childhood programs has contributed to the creation and authenticity of the HighScope Preschool Curriculum over the decades. I hope this book continues to support their learning in science and technology for many years to come.

CHAPTER 1

The Importance of Science and Technology

What Is Science and Technology?

Adults don't have to entice children to become interested in science. Their interest is as natural as the world they are exploring. **Science and technology** builds on children's sense of wonder and curiosity to help them understand how the world works. As they engage in hands-on and multisensory scientific experiences, preschoolers use their emerging abilities in observation, communication, representation, and reasoning to explore the world and share their discoveries. Just like other "academic" areas of early development, science in the preschool years is not about memorizing facts. Rather it entails being actively and directly engaged with objects and events in the real world. By providing appropriate hands-on materials and experiences, and sharing children's delight in discovery, adults support children's early learning in science and technology. A child such as Annie observes and classifies, Karl makes predictions, Noah draws conclusions, Carlos and Nora communicate ideas, and Felix reflects on the natural and physical world (see "Science and Technology in Action," right). Others attempt to explain how tools work, and use technology to choose and learn lessons of interest to them.

Early learning in science and technology is based on a model of scientific inquiry that involves asking and answering questions and applying problem-solving strategies. The emphasis is on the investigative process, although there is also specific content about the natural and physical world that is appropriate and interesting for preschoolers to learn. While young children do not follow the rigorous procedures of adult scientists, they engage in what we call the "preschool scientific method." They try out things with materials and actions that pique their curiosity, observe what happens, attempt

Science and Technology in Action

At planning time, Annie sees a magnetic puzzle in the toy area and says, "This a magnet. It sticks to metal." She tries to make it stick to the clip of the clipboard and observes to her teacher, "But this metal is not sticking."

At work time in the art area, Karl tells Zora it will take a long time for her yarn-and-glue picture to dry because she used a lot of glue. He adds, "If you leave it out all night, maybe the next day, you can take it home." When the teacher asks why, Karl says, "Because it will dry."

When signing in at greeting time, Noah notices two short pencils. "This one is the shortest," he says, "Do you want to know my hypothesis? It's smaller because someone broke it."

At snacktime, while the teacher reads the book Go Dog, Go! *the children discuss why the car one dog is riding in has stopped. Carlos says it's because there is something wrong with the car. "It's smoking because the tire is broken," says Nora. "No," says Carlos, "it's the engine!"*

At snacktime, Felix talks about a story he and his mom read together: "When the squirrel forgets the acorn, it turns into a tree. That's my favorite part."

to make sense of the results, and eagerly share their observations and conclusions — accurate or not — with others. By engaging in these processes, preschoolers are developing their critical-thinking and problem-solving capacities. They use their senses and emerging language and literacy skills, represent their findings in creative ways, and collaborate with one another in their investigations.

Components of Science and Technology

Young children are continually immersed in science "by which we mean the process of developing explanations about how the world works — and are developing useful and quite powerful scientific theories" (Landry & Forman, 1999, p. 133). Children come to science eager to learn about the natural and physical world, just as they are inherently motivated to develop physically and to engage in social interaction. Science fits with the way young children process and make sense of their observations and experiences. Each encounter with the world constitutes a scientific "event" that piques their curiosity and leads to a growing understanding of scientific principles. Science educator Lucia French (2004) describes the process this way:

> During their first exposure to one of these events, children may simply be interested and perhaps surprised. During the second exposure, they are creating a richer representation of similarities and differences across the two experiences. After several exposures, they have created a generalized understanding of that particular aspect of "how the world works" in that particular situation and freely make predictions about "what will happen next" or "what will happen if…." (p. 140)

> "Children are naturally curious about the world and want to find out as much as they can. They want to know what makes the wind blow, how trees grow, why fish have fins, and where turtles go in the winter. But they don't want adults to give them the answers…. They don't want science to be something that is imparted to them; they want it to be something they do."
>
> — Wilson (2002, Conclusion)

The investigative process. As illustrated by the earlier descriptions of young children, science in the preschool years is not about memorizing facts. Neither is it a subject learned in isolation nor confined to a dusty bird's nest or mysterious "pod" on the science table (Neill, 2008). Rather, science is an investigative process that involves observing, predicting, experimenting, verifying, and explaining — activities of interest to and within the capacity of young children. Educators are therefore increasingly introducing science as an integral component of the early childhood curriculum.

Young children's fascination with science spills over into virtually every other area of learning. Engaging in the scientific process develops *critical thinking skills* because children attempt to understand the what, how, and why of the events they observe. Using all their senses to fully experience the world around them heightens their *perceptual abilities*. Scientific inquiry promotes *language* growth because preschoolers are eager to communicate their experiences and share the conclusions they draw. Science learning also builds on and contributes to children's knowledge and skills in *mathematics,* as they count, measure, and look for patterns in the phenomena they observe. Representing

Why Science Belongs in the Early Childhood Curriculum

Curriculum developers and researchers Rochel Gelman and Kimberly Brenneman state that "to do science is to predict, measure, count, record, date one's work, collaborate, and communicate" (2004, p. 156). Similarly, early childhood specialists Carol Seefeldt and Alice Galper define science as the process of "manipulating, observing, thinking, and reflecting on actions and events" (2002, p. 43). These are all abilities that emerge in the preschool years and that early childhood educators strive to promote. Developing competence in these areas also supports children's overall cognitive development, language and literacy skills, and social interactions.

The very expression "to do science" suggests that learning in this area is an active process, rather than the passive memorization of facts. While there is specific knowledge that preschoolers are eager to acquire (such as the names and characteristics of plants and animals), it is hands-on investigation that sustains their interest and curiosity. Scientific inquiry is for inquiring minds. The National Committee on Science Education Standards and Assessment says,

> Science inquiry refers to the diverse ways in which scientists study the natural world and propose explanations based on the evidence derived from the work. Inquiry also refers to the activities of students in which they develop knowledge and understanding of scientific ideas, as well as an understanding of how scientists study the natural world. (1996, p. 23)

Professors Haim Eshach and Michael Fried (2005) assert that young children should be exposed to science for the following six reasons:

- Children naturally enjoy observing and thinking about nature.
- Early exposure develops positive attitudes toward science.
- Early scientific experiences form the basis for later formal education.
- Using scientific language at a young age helps to develop scientific concepts.
- Young children are beginning to reason scientifically.
- Science experiences help to develop scientific thinking about the world.

Given the affinity between children and scientific inquiry and learning, science education is a natural and necessary component of the early childhood curriculum.

their experiences and ideas, such as drawing at recall time the pulley they played with at work time, engages children in *creative arts*. Finally, scientific inquiry involves *social collaboration,* for example, when two children fill the buckets on either side of a balance scale with blocks of different sizes. In fact, conflicting explanations by their peers more often motivates children to change their mathematical and scientific theories than do comments or questions from adults (Tudge & Caruso, 1988; Vygotsky, 1978).

The development of scientific thinking. The ability to "do" science relies on a progression in how children think about the world, and involves three interrelated aspects of early cognitive development. The first ingredient is recognizing that something that was expected to happen did not occur. Very young children tend to simply accept what they see, hear, and feel. Everything in the world is new and a mystery to them, so they accept discoveries without question. However, as they accumulate knowledge, children set up expectations based on what they've already learned. As science researchers Rochel Gelman and Kimberly Brenneman (2004) note, "A key finding from cognitive psychology…is that it is easier to learn more about what one already knows than to build concepts in a new domain about which one has little or no relevant knowledge" (p. 155). Put another way, we could say that "the more you know, the more you notice." Consider the following examples:

At work time at the sand table, Minyi says excitedly, "Look, there's something shiny buried down there."

At outside time, while Josh is swinging, he says, "When I go forward, the sand is behind me. When I go backward, the sand is in front of me. It's like we're both changing directions."

At work time in the house area, Ayin observes, "If I comb her [the doll's] hair, it makes it smooth."

At outside time, Rennie runs to his teacher and says, "The bird in the birdhouse has yellow feathers in its tail!" Saul coming up behind him says, "No. They're green." The two boys and their teacher walk to the birdhouse slowly, "so they won't scare away the bird," and look at it together. After some discussion, the boys agree the feathers are "green with lots of yellow."

At small-group time, Matthew mixes watercolor paints. He tells his teacher, "I'm sciencing." When she asks what that means, he says, "Sciencing is when you are figuring something out."

Thus, when something unexpected happens, young children ask, *"What's wrong here?"* Because of the connections they've made from previous experiences, preschoolers do not treat

Science in the preschool classroom is not about memorizing facts but rather involves children's active engagement with objects and events in the real world.

each thing that happens as a singular event. They wonder if it represents a more general principle (Landry & Forman, 1999). For example, instead of merely observing that a Popsicle stick floats, a child may notice that other things, such as rocks, sink. From this the child draws the conclusion that light things float and heavy things sink. However, when something he or she expects to float instead sinks, such as a plastic rod the same size and shape as the stick, the child has to resolve this discrepancy. Once the child meets with an unexpected result, it spurs his or her scientific inquiry.

Science and Technology

With adult support and encouragement, preschool children can engage in more systematic investigations to test their ideas, create explanations, and draw conclusions.

At work time in the art area, Nell covers her entire paper with paint. When she removes it from the easel, she notices two white spots where it was attached with clothespins. The previous times Nell painted, she did it on the table where nothing blocked the paint from the paper. After unpinning the painting, she sets it on the table and fills in the spaces before clipping it on the line to dry.

❖

At work time in the art area, Justin shakes and squeezes the glue bottle but nothing comes out. Lyle, who is working beside him, explains that "It's stuck." Lyle gets a nail from the woodworking area and pokes it in the tip of the glue bottle. "There," he demonstrates to Justin, "now it's not stuck anymore." As Justin continues to work, he sticks the nail in the tip of the glue bottle before he uses it each time "so it won't get stuck again."

❖

At outside time, Gretchen notices that some worms are crawling and others are curled up. "Why do you suppose that is?" asks her teacher.

Gretchen watches them a while and then concludes, "These worms are all curled up because they're still asleep."

Recognizing the need to adjust their thinking, preschoolers next ask, *"What is happening here?"* They may first try to repeat the situation to make sure their initial observation was accurate. Next they may try slight variations. For example, they might push and hold the stick underwater for a while before letting go to see if it stays there. Or they might lay the plastic rod very gently on top of the water to see if it floats. They may even get a heavier wooden block to see if it sinks or floats. Based on their observations, young children are motivated to construct an alternate theory. It may not be accurate. For example, if both the rock and the plastic rod happen to be colored, they may decide that painted things sink while unpainted things (plain wooden sticks and blocks) float. However, if the theory "fits" within their experience, children will be satisfied.

The final step in engaging in scientific inquiry is answering the question *"Where's the proof?"* On their own, preschoolers are not likely to methodically test their theories with a variety of materials or in different situations. They depend on intuition and reason by analogy (Landry & Forman, 1999). Thus, if the red rock and blue rod sink, while natural wood items do not, it must mean that only colored objects sink. However, with adult support, preschoolers can engage in more systematic investigations and refine their explanations to accommodate what they observe. In these situations, the role of the adult is *not* to tell children they are wrong or correct their ideas. Rather, adults provide appropriate materials and support "so children can test and find out on their own whether their ideas are correct or not" (DeVries & Sales, 2011, p. 2). Thus, an adult hearing the children's theory about floating and sinking might supply colored blocks and clear plastic rods for them to test. Using open-ended questions ("What do you see?" "Why do you think that happened?"), adults can encourage children to create explanations, make predictions based on their theories, observe the outcome of their experiments, and draw new conclusions.

At work time in the block area, Mo builds an increasingly long magnet train to see how many he can pull across the bridge he's made with blocks. He pulls four successfully, but when he adds a fifth magnet to his train, it falls off. He tries again and this time it works with five, so he adds a sixth magnet which falls off and pulls the fifth one off with it. For the rest of work time, Mo pulls a four-magnet train across the bridge and the other structures he builds (apparently concluding that five or more magnet-trains are likely to fall).

At work time at the water table, Salima takes out all the items that float and sets them aside in a bowl. "I only want the ones that sink," she tells her teacher, "because I'm hiding buried treasure."

At outside time, Ben and Julia jump into a pile of leaves. "We need more," Julia says, "to make it soft on the bottom!" They pile on more leaves and jump on top. "Nope. We still need more," says Julia. They continue to add leaves until Julia decides there are enough. Then Ben and Julia spend the rest of outside time jumping into their "supersoft" pile of leaves.

Sharing scientific ideas. As their abilities to engage in scientific exploration unfold, children also become increasingly able to represent and communicate their ideas. They apply their

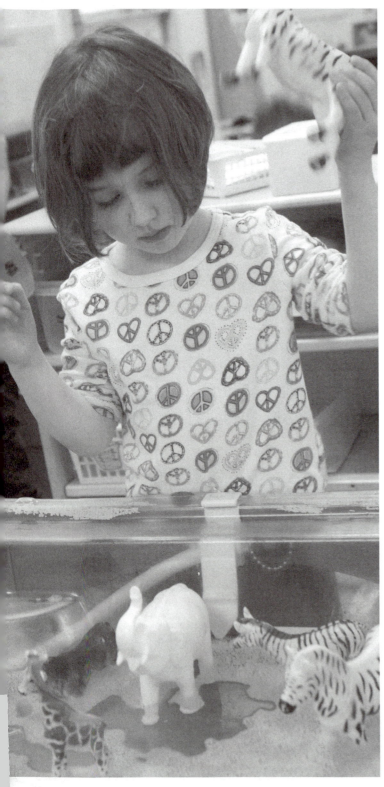

When children sort, compare, and order things in terms of their observable characteristics, they are "doing science."

emerging skills in mathematics (data analysis) to science. For example, to support preschoolers who are interested in classifying objects into those that sink and those that float, a teacher might help children make a simple two-column chart to describe and compare the properties of the two sets of objects. The children may (or may not) come to the conclusion that the type of material affects the buoyancy of an object; for example, wood and air-filled items float, while stone and solid items sink. Adults can also help children learn the vocabulary to label the categories they create, describe their observations, and explain their reasoning.

At greeting time, Jibreel points out the window and says, "Hey, it's snowing. It's winter." He asks his teacher if she will read him The Snowy Day *at work time.*

At outside time, Lucy kneels on the ground to look at the falling snowflakes with a magnifying glass. At work time that afternoon, she glues pieces of white yarn onto construction paper and tells her teacher she is making a "snowflake painting." Her teacher listens as Lucy describes how her painting resembles what she saw under the magnifying glass: "The snowflake had lots of arms coming out from the middle, like these pieces of yarn."

❖

At work time in the house area, Marlis tells her teacher, "My mommy uses a round pan like that when she bakes my birthday cake. It's better than the flat pan. That's for cookies." The teacher wonders aloud what can be cooked in the other pots and pans in the house area, which differ in size and shape. Marlis points to a sauce pan and says, "You can cook soup in this one."

By providing children with appropriate hands-on materials and experiences, and sharing children's delight in discovery, adults can promote active science learning throughout the program day.

Activities in "doing science." Because learning about how the world works is essentially what young children do all the time, science and technology education occurs all the time in an early childhood setting. While this is true, it is nevertheless helpful to identify the particular activities that constitute "doing science" at this age so that adults can intentionally and effectively support learning in this content area. Integrating the work of several curriculum developers and researchers (French, 2004; Gelman & Brenneman, 2004; Worth & Grollman, 2003), we can say that science learning happens when children do the following:

- Use all their senses to explore objects and events
- Ask questions about objects and events in their environment
- Act upon things and observe what happens
- Describe, sort, compare, and order things in terms of observable characteristics
- Use simple tools to carry out and extend their observations
- Engage in simple investigations by making predictions, experimenting, gathering data, recognizing patterns, and drawing conclusions
- Represent and record their observations
- Work collaboratively
- Communicate their ideas, explanations, and theories
- Listen to new perspectives and adjust their ideas accordingly

Providing young children with opportunities to engage in these activities ensures that their natural interest in science will not only thrive, but also result in exciting and meaningful learning.

About This Book

In the HighScope Preschool Curriculum, the content of children's learning is organized into eight areas: A. Approaches to Learning; B. Social and Emotional Development; C. Physical Development and Health; D. Language, Literacy, and Communication; E. Mathematics; F. Creative Arts; G. Science and Technology; and H. Social Studies. Within each content area, HighScope identifies **key developmental indicators (KDIs)** that are the building blocks of young children's thinking and reasoning.

The term *key developmental indicators* encapsulates HighScope's approach to early education. The word *key* refers to the fact that these are the meaningful ideas children should learn and experience. The second part of the term — *developmental* — conveys the idea that learning is gradual and cumulative. Learning follows a sequence, generally moving from simple to more complex knowledge and skills. Finally, we chose the term *indicators* to emphasize that educators need evidence that children are developing the knowledge, skills, and understanding considered important for school and life readiness. To plan appropriately for students and to evaluate program effectiveness, we need observable indicators of our impact on children.

This book is designed to help you as you guide and support young children's learning in the Science and Technology content area in the HighScope Curriculum. This chapter provides insights from the research literature on the development of scientific thinking in young children and summarizes the steps in their acquisition of scientific knowledge and principles. Chapter 2 describes general teaching strategies for Science and Technology and provides an overview of the KDIs for this content area.

Chapters 3–10, respectively, provide specific teaching strategies for each of the eight KDIs in Science and Technology:

45. **Observing:** Children observe the materials and processes in their environment.

46. **Classifying:** Children classify materials, actions, people, and events.

47. **Experimenting:** Children experiment to test their ideas.

48. **Predicting:** Children predict what they expect will happen.

49. **Drawing conclusions:** Children draw conclusions based on their experiences and observations.

50. **Communicating ideas:** Children communicate their ideas about the characteristics of things and how they work.

51. **Natural and physical world:** Children gather knowledge about the natural and physical world.

52. **Tools and technology:** Children explore and use tools and technology.

At the end of each of these chapters is a chart showing ideas for scaffolding learning for that KDI. The chart will help you recognize the specific abilities that are developing at earlier, middle, and later stages of development and gives corresponding teaching strategies that adults can use to support and gently extend children's learning at each stage.

The Importance of Science and Technology 11

Because young children are always exploring and learning about the world and how it works, science and technology education occurs all the time in an early childhood setting. Adults identify the kinds of activities that constitute "doing science" at this age in order to intentionally and effectively support children's learning in this content area.

HighScope Preschool Curriculum Content
Key Developmental Indicators

A. Approaches to Learning

1. **Initiative:** Children demonstrate initiative as they explore their world.
2. **Planning:** Children make plans and follow through on their intentions.
3. **Engagement:** Children focus on activities that interest them.
4. **Problem solving:** Children solve problems encountered in play.
5. **Use of resources:** Children gather information and formulate ideas about their world.
6. **Reflection:** Children reflect on their experiences.

B. Social and Emotional Development

7. **Self-identity:** Children have a positive self-identity.
8. **Sense of competence:** Children feel they are competent.
9. **Emotions:** Children recognize, label, and regulate their feelings.
10. **Empathy:** Children demonstrate empathy toward others.
11. **Community:** Children participate in the community of the classroom.
12. **Building relationships:** Children build relationships with other children and adults.
13. **Cooperative play:** Children engage in cooperative play.
14. **Moral development:** Children develop an internal sense of right and wrong.
15. **Conflict resolution:** Children resolve social conflicts.

C. Physical Development and Health

16. **Gross-motor skills:** Children demonstrate strength, flexibility, balance, and timing in using their large muscles.
17. **Fine-motor skills:** Children demonstrate dexterity and hand-eye coordination in using their small muscles.
18. **Body awareness:** Children know about their bodies and how to navigate them in space.
19. **Personal care:** Children carry out personal care routines on their own.
20. **Healthy behavior:** Children engage in healthy practices.

D. Language, Literacy, and Communication[1]

21. **Comprehension:** Children understand language.
22. **Speaking:** Children express themselves using language.
23. **Vocabulary:** Children understand and use a variety of words and phrases.
24. **Phonological awareness:** Children identify distinct sounds in spoken language.
25. **Alphabetic knowledge:** Children identify letter names and their sounds.
26. **Reading:** Children read for pleasure and information.
27. **Concepts about print:** Children demonstrate knowledge about environmental print.
28. **Book knowledge:** Children demonstrate knowledge about books.
29. **Writing:** Children write for many different purposes.
30. **English language learning:** (If applicable) Children use English and their home language(s) (including sign language).

---

[1] Language, Literacy, and Communication KDIs 21–29 may be used for the child's home language(s) as well as English. KDI 30 refers specifically to English language learning.

E. Mathematics

31. **Number words and symbols:** Children recognize and use number words and symbols.
32. **Counting:** Children count things.
33. **Part-whole relationships:** Children combine and separate quantities of objects.
34. **Shapes:** Children identify, name, and describe shapes.
35. **Spatial awareness:** Children recognize spatial relationships among people and objects.
36. **Measuring:** Children measure to describe, compare, and order things.
37. **Unit:** Children understand and use the concept of unit.
38. **Patterns:** Children identify, describe, copy, complete, and create patterns.
39. **Data analysis:** Children use information about quantity to draw conclusions, make decisions, and solve problems.

F. Creative Arts

40. **Art:** Children express and represent what they observe, think, imagine, and feel through two- and three-dimensional art.
41. **Music:** Children express and represent what they observe, think, imagine, and feel through music.
42. **Movement:** Children express and represent what they observe, think, imagine, and feel through movement.
43. **Pretend play:** Children express and represent what they observe, think, imagine, and feel through pretend play.
44. **Appreciating the arts:** Children appreciate the creative arts.

G. Science and Technology

45. **Observing:** Children observe the materials and processes in their environment.
46. **Classifying:** Children classify materials, actions, people, and events.
47. **Experimenting:** Children experiment to test their ideas.
48. **Predicting:** Children predict what they expect will happen.
49. **Drawing conclusions:** Children draw conclusions based on their experiences and observations.
50. **Communicating ideas:** Children communicate their ideas about the characteristics of things and how they work.
51. **Natural and physical world:** Children gather knowledge about the natural and physical world.
52. **Tools and technology:** Children explore and use tools and technology.

H. Social Studies

53. **Diversity:** Children understand that people have diverse characteristics, interests, and abilities.
54. **Community roles:** Children recognize that people have different roles and functions in the community.
55. **Decision making:** Children participate in making classroom decisions.
56. **Geography:** Children recognize and interpret features and locations in their environment.
57. **History:** Children understand past, present, and future.
58. **Ecology:** Children understand the importance of taking care of their environment.

CHAPTER 2

General Teaching Strategies for Science and Technology

A great deal of early science learning takes place spontaneously and informally. Children engage in science when they pose and answer "how, what, and why" questions, resolve discrepancies between what they expect to happen and what they actually observe, and solve problems using and transforming materials. However, while preschoolers are "natural scientists," that does not mean early science learning can be left to chance. Adults must plan activities and build on unplanned discoveries to help young children develop the habits of mind and particular skills associated with scientific inquiry. Children also depend on adults to introduce them to specific knowledge and simple principles in biology, chemistry, and physics (for example, that plants grow from seeds; that heat transforms the appearance or texture of some materials). To be an active participant in preschoolers' scientific explorations, adults can use the general strategies listed in this chapter.

When children engage in the steps of the "preschool scientific method," they use the same processes as adult scientists, though their investigations are more spontaneous and wide ranging.

General Teaching Strategies

Introduce children to the steps in the scientific method

Just as preschoolers enjoy thinking of themselves as readers or artists, they also feel satisfaction in seeing themselves as scientists. Young children use the same processes as adult scientists, but HighScope refers to this as the *"preschool scientific method* [because] preschoolers engage in their investigations in a much more random and spontaneous fashion" (Neill, 2008, p. 2)[2]. These steps are not a rigid set of procedures, but rather a wide-ranging exploration of the world and how it works. Hence, scientific investigation often occurs while learning about other subjects.

Since "doing science" is a process, you can help children become familiar with and carry out each step in the "preschool scientific method." Attach simple labels to children's activities, such as describing properties, making predictions, conducting experiments, recording data, and explaining outcomes. Preschoolers spontaneously undertake these activities but, as previously noted, not necessarily in a systematic way. If we are serious about closing the "science gap" and starting more young people on the path to becoming scientists, we can begin by encouraging preschoolers to see themselves as capable of going through the same series of steps as grown-up scientists.

To achieve this goal, Rochel Gelman and Kimberly Brenneman (2004) suggest that teachers introduce preschoolers to "the vocabulary and methods of *observe, predict,* and *check*" (p. 153). They give an example in which children used all their senses to explore a whole apple and the teacher writes down what they say ("The apple is red, round, smooth, and cold") "because

---

[2]The six components of the preschool scientific method comprise the first six science KDIs, described later in this chapter.

scientists record their observations." Next children were asked to make a prediction, "something like a guess," about what was inside the apple ("It's white, it has seeds"). Finally, after they cut open their apples, children checked their predictions against what they found (in addition to being white and having seeds, "It's wet").

Here's another example from a HighScope classroom:

At work time in the art area, Jamal makes an all-glue picture and asks his teacher to write on a sticky note, "This is a glue picture. It is white and sticky." He wants to take the picture home but observes that the glue is still very wet and he can't carry it without dripping. When the teacher asks Jamal what he can do about the problem, Jamal predicts that the glue will be dry by the next day. The following morning, Jamal checks to see whether the glue has dried. He sees and feels that it has, and puts the picture in his cubby to take home.

Encourage reflection

Although children have inquiring minds, their curiosity is often satisfied by taking action and observing what happens. Science, however, also means explaining the results. As noted above, while preschoolers are expanding their knowledge base, much of the world is still new to them. Therefore, they may simply accept what they observe as the way things are, without asking or attempting to answer "how" or "why" questions on their own. It is up to teachers to establish a reflective spirit in the classroom that

Adults create openings for conversation and comments that encourage reflection when they work alongside children and share their discoveries.

encourages young children to adopt these inquiring habits of mind. As science educators Karen Worth and Sharon Grollman (2003) emphasize,

> Direct experience with materials is critical but it is not enough. Children also need to reflect on their work. They need to analyze their experiences, think about ideas such as patterns and relationships, try out new theories, and communicate with others. These processes allow children to think in new ways about what they did, how they did it, and what is significant to them. (p. 5)

The HighScope plan-do-review cycle establishes a general climate and specific time period to support reflection in the classroom. However, adults also need to help children focus and reflect on their scientific explorations in particular. Working alongside children and sharing in their discoveries creates openings for you to make conversational comments that encourage reflection, such as "Show (tell) me how you did that," "How could we make it happen again?" "I wonder why that happened," or "What do you suppose would happen if…?" If children get stuck, you can also help by restating the question or problem in a way they can answer. For example, you might say, "Let's see if there's something the same about these two things that makes them different from that one."

Create opportunities for surprise and discrepancy

Young children's ideas about science develop when they encounter the unexpected, that is, when something they anticipate turns out differently. So one way to encourage the growth of thinking and reasoning skills in science is to create opportunities where children are likely to experience "cognitive conflict," that is, discrepancies between what they believe and what they observe.

Opportunities for science learning exist "here, there, and everywhere" — in the classroom and on the playground!

You can create natural (not contrived) opportunities in many ways. Simply providing many types of materials and tools for children to explore guarantees they will encounter surprises by chance. However, you can also be intentional about providing new materials that behave differently from familiar ones, or experiences in which familiar materials behave in different ways. For example, if children have been building with regularly shaped blocks, introduce blocks that are irregular in shape so they can explore physical properties related to balance and fit. If children have been working with clay inside the classroom, bring the clay outside where the sun makes it dry faster. If children have been racing cars down wood or plastic ramps, do a small-group time in which they work with ramps covered in felt, sand paper, and other smooth and rough textures.

Merely encountering the unexpected may not prompt children to seek alternative explanations, however (Chinn & Schaverien, 1996). They might simply call whatever happens "magic." For surprises to encourage children to re-examine their own thinking, a spirit of reflection (discussed earlier) must first prevail in the classroom. Children will be less likely to regard the discrepancy as magic, and will instead wonder why their ideas do not hold up.

Once children accept that an alternative reason is needed, you can guide them to consider other possibilities. Simply telling them the correct answer or giving them a reason is unlikely to change their thinking. However, providing children with a variety of related materials, encouraging them to try out their own ideas, and occasionally suggesting other ways they might test their theories will open the door to self-discovery. What the children then observe and conclude on their own — whether or not they are correct — engages them in scientific rather than magical reasoning. As Landry & Forman (1999) point out, "Unexpected events, when embedded in the child's own processes of exploration, provide rich opportunities for encouraging the flow of thinking" (p. 147).

Encourage documentation

It takes time and repeated instances for children to change their ideas. Preschoolers are likely to focus on what they see and hear in the moment, and their memories are fragile (they forget things); they may therefore not recall how something they observed a day or even ten minutes ago contradicts their current reasoning. Repeated examples can help fix what they observe in memory. Also, although they still mostly operate at a concrete level, three- and four-year olds are capable of creating and interpreting representations to help them document and remember things. This allows them, with adult assistance, to record data by drawing pictures and interpreting simple charts, graphs, and photos (Katz & Chard, 1996). Documentation serves as a visual reminder as children construct their explanations. In addition to supporting their scientific reasoning, recording information builds on their emerging mathematical (data analysis) skills.

Encourage collaborative investigation and problem solving

Collaboration has long been recognized as a vital component in the development of social skills. However, collaborative problem solving is also an important factor in young children's cognitive development (Tudge & Caruso, 1988). Working with peers is especially effective in advancing early mathematical and scientific thinking. Social interactions don't just help children construct and change their ideas, it actually deepens their understanding.

At a simple level, the more eyes and ears are involved, the greater the volume of data the children can collect. Each child observes things that the others do not. Because they have to describe and attempt to explain what they see, working cooperatively enhances children's language skills. Language complexity and thinking complexity are connected, both in the number of details children observe and report, and in the way they relate these elements to one another.

Also, as noted earlier, children are more influenced to change their thinking when challenges come from peers rather than from adults. Perhaps this is because they are used to adults seeing things differently, but when someone who is like them has an alternative perception

Science and Technology in Action

KDI 45. Observing

KDI 46. Classifying

KDI 47. Experimenting

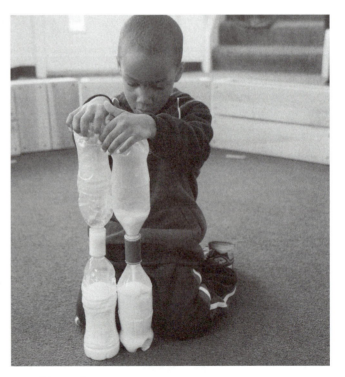

KDI 48. Predicting

General Teaching Strategies for Science and Technology 21

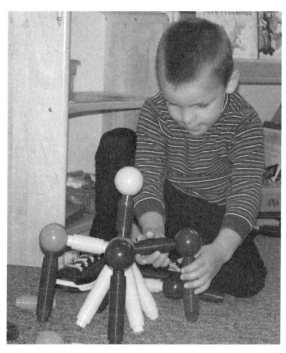

KDI 49. Drawing conclusions

KDI 50. Communicating ideas

KDI 51. Natural and physical world

KDI 52. Tools and technology

or explanation, they have more reason to question their own thinking. Children may also be socially motivated to change their cognitive concepts. Just as they want to resolve social conflicts and be accepted by peers, young children may be open to altering their scientific beliefs based on their desire to go along with the group. But while children might at first agree simply for the sake of being agreeable, in a classroom where reflection and authentic respect for others' ideas are the norm, young children are capable of genuinely examining and gradually changing their thinking.

At small-group time, the children plant tomato seedlings in their vegetable garden. Fiona and Gabriella disagree over how much water to use. Fiona fills half a pail, but Gabriella, who has planted a garden at home with her mother, says it's too much. When Fiona pours the water on a tomato plant, she sees it lean to the side, while the one Gabriella put less water on is still standing upright. Fiona does not say anything, but for the next seedling she uses less water.

Opportunities for science learning in the preschool classroom really do exist "here, there, and everywhere" (Neill, 2008). All you need to do is use your "science eyes" and you will find them. Moreover, as you implement the general strategies described earlier and the specific ones described in the following chapters, be sure to share the children's sense of wonder about the discoveries you make together.

Key Developmental Indicators

HighScope has eight **key developmental indicators (KDIs)** in the content area of Science and Technology: 45. Observing, 46. Classifying, 47. Experimenting, 48. Predicting, 49. Drawing conclusions, 50. Communicating ideas, 51. Natural and physical world, and 52. Tools and technology. The first six topics refer to the processes or "preschool scientific method" that young children use to investigate their universe (Neill, 2008). The other two topics deal with the kinds of knowledge and skills they acquire through their explorations.

Chapters 3–10 discuss knowledge and skills young children acquire in each of these KDIs and the specific teaching strategies adults can use to support their development. At the end of each chapter is a "scaffolding chart" with examples of what children might say and do at early, middle, and later stages of development, and how adults can scaffold their learning through appropriate support and gentle extensions. The chart offers additional ideas on how you might carry out the suggested strategies during play and other interactions with children.

Key Developmental Indicators in Science and Technology

G. Science and Technology

45. Observing: Children observe the materials and processes in their environment.

Description: Children are curious and use all their senses to learn more about the natural and physical world. They gather information by observing what others do and discovering how tools and materials work.

46. Classifying: Children classify materials, actions, people, and events.

Description: Children group similar things together. They identify relationships between things and the categories they belong to. Children look for new ways to organize the knowledge they already have and for ways to fit new discoveries into familiar categories.

47. Experimenting: Children experiment to test their ideas.

Description: Children experiment to test whether an idea is true or a solution will work. They may encounter problems with materials that they do not have answers for. They experiment by manipulating materials, using trial and error, and then approaching the problem with possible solutions in mind.

48. Predicting: Children predict what they expect will happen.

Description: Children indicate through words and/or actions what they expect an outcome to be. They think about what happened in similar situations and anticipate what might happen. Children make predictions based on experimentation.

49. Drawing conclusions: Children draw conclusions based on their experiences and observations.

Description: Children attempt to fit their observations and reasoning into their existing knowledge and understanding. They construct knowledge in their own way as they collect data to help them form theories about how the world works (e.g., "It's night because the sun goes to bed").

50. Communicating ideas: Children communicate their ideas about the characteristics of things and how they work.

Description: Children share their questions, observations, investigations, predictions, and conclusions. They talk about, demonstrate, and represent what they experience and think. They express their interest in and wonder about the world.

51. Natural and physical world: Children gather knowledge about the natural and physical world.

Description: Children become familiar with characteristics and processes in the natural and physical world (e.g., characteristics of plants and animals, ramps and rocks; processes of growth and death, freezing and melting). They explore change, transformation, and cause and effect. They become aware of cycles that are meaningful to them.

52. Tools and technology: Children explore and use tools and technology.

Description: Children become familiar with tools and technology in their everyday environment (e.g., stapler, pliers, computer). They understand the functions of equipment and use it with safety and care. They use tools and technology to support their play.

CHAPTER 3

KDI 45. Observing

G. Science and Technology
45. Observing: Children observe the materials and processes in their environment.

Description: Children are curious and use all their senses to learn more about the natural and physical world. They gather information by observing what others do and discovering how tools and materials work.

At small-group time, while using nuts and bolts, Dwight says, "Hey look, this nut fits on this bolt."

At outside time, Noni and Paul squat to look at bugs exposed when the custodian turned over a rock. Noni observes, "It's got wavy things on top of its head." Paul adds, "And tickly feet."

At outside time, Robby points to a dry area under the slide after it has rained and says to his teacher, "This spot is still light." His teacher suggests they feel the light spot and a dark spot next to it. "It's dry!" Robby declares in amazement, "and this spot is wet." "I wonder why," says his teacher. Robby points to the wet area and explains, "The cloud stayed over there."

At pickup time, Belinda pulls her mother over to look at a vine. "There are leaves growing out of the ground, up the tree, and past the crow's nest," she points out to her mother.

How Observing Develops

Unlike casual looking, observing is "paying close attention to something to learn more about it" (Neill, 2008, p. 10). It is an intentional action that involves looking with purpose. It is the first step in collecting information and making sense of how the world works. Being a careful and accurate observer is the single most important ability in becoming a good scientist.

Young children depend on their senses to gather data about the natural and physical world. They learn how things appear, sound, feel, smell, and taste. They discover what things can and cannot do, and see the effects of their own actions, the actions of others, and the actions of one object upon another. Young children also begin to notice intangible factors in the environment, for example, light (how bright something appears) and heat (how fast something melts).

Although most preschoolers are curious and eager to explore using all their senses, some may be reluctant or cautious about doing so due to innate temperamental differences (Chess & Thomas, 1996) or experiences outside of school. Remember to respect each child's preferences and create a trusting environment in which the children choose what to explore and how to make their observations.

Children's interest in and ability to observe an increasing number of details also develop over time. As in many other content areas, the growth of observational skills proceeds from the few to the many, the simple to the complex, and the isolated to the connected. Thus younger children tend to focus on a few familiar characteristics, such as the color of a block or how fast a car goes down the ramp. Older preschoolers, by contrast, may differentiate among color gradations (light, medium, and dark red) and several speeds (slow, faster, even faster, and fastest). They will also observe how their actions affect the properties of objects and how they behave (e.g., varying the ramp's angle affects speed).

Finally, related to the increasing level of detail and the interrelationships they perceive, children also develop a growing ability to describe their sensory impressions and the properties of objects and actions. Their observations progress from being predominantly physical to becoming more verbal (Grunland, 2006). Initially, they may simply experience the sensations or use their bodies and materials to observe the effects they produce. As their observational skills and vocabularies grow, they attach simple words to describe what they see, hear, feel, taste, and smell. Later still, children are eager to describe in their own words what they observe, what they think, and how they feel about the world they are discovering.

Teaching Strategies That Support Observing

Although young children are naturally inclined to observe the world around them, they depend on adults to provide an environment with objects and events worthy of their focused attention. To support and enhance preschoolers' observational skills, teachers can use the following strategies.

Provide a sensory-rich environment

At work time in the house area, Gina stirs pebbles in a bowl to make "stew." "These look like peas," she tells her teacher, "but they are too hard to really eat."

On the way to the park for outside time, Tucker and Brandon look forward to walking past the entrance to the parking garage so they can call inside and make echoes.

At work time in the block area, Chelsea hears someone come in the preschool door. "It's Corrin," she says, "I can hear her voice."

At snacktime, Julia takes a sip of juice and says, "This has pineapple in it."

At small-group time at the sensory table, the children explore bubble bath and share their observations. Jayla says it looks and smells like shampoo. Sloan immerses his hands in the liquid and says it feels "slimy," while Zak describes the texture as "gloopy." Carla declares that the bubble bath "feels like glue, looks like glue, and smells like strawberries."

At work time in the art area, Jorge cuts along one edge of the paper with plain scissors and another edge with serrated scissors. He runs his finger along each edge of the paper and says, "I like the bumpy one better." He uses serrated scissors to cut the remaining two edges.

An indoor and outdoor learning environment with diverse sensory experiences encourages children to observe their surroundings using all the perceptual capacities they are born with. To help both you and them be intentional about actively observing materials and processes, set up a "sensory table" (available through school supply retailers) or simply use galvanized metal tubs or large plastic basins. Fill the table or basin with a material such as sand, water, shaving cream, dishwashing liquid, beads, cotton balls, pebbles, bark, acorns, or shredded paper. Periodically change the filler so children can explore and discover the properties of a variety of materials.

In addition to the sensory table, include appealing materials that children can observe with all their senses throughout the indoor and

An indoor and outdoor environment that appeals to children's senses will help to develop their observational skills. A "sensory table" can encourage children — and their teachers — to become more intentional about actively observing materials and processes.

outdoor learning environment. For example, provide things that create light and shadow (flashlights, cellophane taped to windows, flying banners, wind spinners, sheets hung up or draped over tables and chairs); distinctive textures (bark, straw, gourds, cotton, leather); aromatic materials (spice jars, beeswax, open windows that let in the smells of plants and rain, herb garden); things that make noise (musical instruments, ticking clocks and timers, workbench tools, pebbles and beads to pour into and out of containers, running water, bird feeders that attract songbirds outdoors); and snack foods with a variety of tastes, smells, and textures (fruits and vegetables, grains, condiments).

Introduce children to new materials at small-group time so they can explore and learn about their sensory properties, and then make them available for use at work time and other times of the day. In her book *Real Science in Preschool,* HighScope early childhood specialist Polly Neill (2008) describes a small-group time in which children are encouraged to observe and describe the properties of three items in their baskets: cotton batting, pine cones, and crumpled aluminum foil. The teacher invites the children to consider how the items *feel* and how they *look,* and makes a chart to write down their discoveries. Children describe the cotton batting as "soft," "like Lamby," "a cloud," and "fluffy." Their observations about the pine cones include "ow," "prickery," "very rough," "bumpy," and "sticky." About the aluminum foil, they observe that it is "noisy," "shiny," and "rough." One child associates the foil with pizza. Let children begin observing materials by using just their senses and bodies. For example, encourage them to feel the water with their hands — patting the surface, submerging up to their fingernail or wrist, splashing gently or with gusto to see how far the water spreads, or wiggling their fingers to create ripples. Later on, add various materials and tools (stirrers, cups, slotted spoons, items that sink or float, paint or dye) so children can experiment and observe additional properties.

Establish a safe environment for children to observe with all their senses

Of equal importance in providing diverse sensory materials is the climate or spirit in which adults offer them. Children need freedom to explore with all their senses and to observe results without sanctions. They should be allowed to take risks, make a mess, and try actions and combinations that don't work. At the same time, children must also have the right to refuse to taste, smell, touch, or otherwise experience any of the materials provided. In other words, children should be given full autonomy over what to explore and how they conduct observations about their environment.

At planning time, Lizzie says she is going to work in the art area. "Yesterday I mixed red and yellow paint. It made orange!" Once in the art area at work time, Lizzie mixes red and blue and pronounces it "purple." However, she decides it is too dark and adds more blue, which only makes it darker. She tries adding yellow, which turns it brown. Lizzie sighs and pours the paint in the sink. "Maybe tomorrow I will make purple," she tells her teacher.

At snacktime, Manola's mother brings in tortillas and salsa. Most of the children eat them, comparing them to pancakes and saying the salsa tastes spicy. One child smells a tortilla but does not eat it, and two shake their heads to indicate they are not interested in trying tortillas. The teacher provides a basket of crackers for children who prefer to eat something familiar.

Children need freedom to explore with all of their senses. They should also be allowed to take risks and make a mess, trying new approaches and combinations in their use of materials.

The same accepting environment that enables children to feel secure expressing themselves emotionally and creatively will also allow them to take risks as they make observations. Children need to trust that adults will not deliberately expose them to harmful or unpleasant sensations. This is not to say children will not occasionally dislike the taste, smell, feel, appearance, or sound of something. Children with special needs may have particular sensory sensitivities, but any child may resist or react to a given sensory experience. Therefore, children should know that their introduction to new sensations will be done gradually and in manageable amounts. They will not be bombarded by overwhelming levels of stimulation or too many sensations at once. Likewise if children want to stop an experience, or don't want to try something at all, they need the security of knowing their choices are acceptable. Materials or events are there if they change their minds, but there should never be any pressure to do so. Children's exploratory observations should always be motivated by their own satisfaction.

Provide the vocabulary to help children label, understand, and use their observations

Support and occasionally extend the language children use as they explore and observe their environment. An ever-growing vocabulary attunes their senses to an increasing array of physical properties and processes. In addition to providing labels for objects (nouns) and actions (verbs), take advantage of descriptive words (adjectives, adverbs) to enhance children's observational skills. Provide, and encourage children to provide, verbal explanations of what they sense ("How does it feel?" "What does that sound remind you of?"), what they do ("How did you get it to stick?" "What should I do to make mine look the same as yours?"), and the outcomes they observe ("I wonder why it got bigger?" "Why do you think it smells more when you crush it?").

To support children's observations and scaffold their learning, match your language to their developmental level. Repeat their words and then label other clearly observable cues like color or loudness. Gradually introduce less obvious properties, such as dampness or temperature. Focus on one or two properties ("It's heavy"; "That feels smooth") and add more as children become familiar with the objects and processes they explore ("The plant is tall and skinny"; "The bird picked up a stick and flew to its nest"). Finally, after focusing on how things

Adults can support and extend the language children use as they explore and observe their environment. As children's vocabulary expands, so does their ability to comprehend and describe their experiences.

look (sound, feel, taste, or smell), talk about how things change based on the children's own actions and those of others, and interactions among the materials themselves ("When the sun came out, it melted the snow on the sidewalk"; "First the cup floated. Then you poured water in it and it sank").

Language affects children's perception and cognition. The bigger children's vocabulary, the greater their ability to comprehend and describe their experiences ("Claire's slippers are furry inside like my coat"). Language also lets them generate their own ideas about how the world works ("If you mix red with white, you get pink"). Giving children the words to describe what their senses register creates the mental categories they can use to "store" their observations. These categories make the information easier to recall, and also render it more available for them to apply in other situations ("My tower didn't fall down when I put the biggest block on the bottom. Maybe I can try the same things with these boxes").

For examples of how young children observe their world at different stages of development and how adults can scaffold their science learning through observation, see "Ideas for Scaffolding KDI 45. Observing" on page 32. Try some of the ideas on the chart, in addition to those described in this chapter, as you play and interact in other ways with the children in your preschool program.

Ideas for Scaffolding KDI 45. Observing

Always support children at their current level and occasionally offer a gentle extension.

Earlier	Middle	Later
Children may	*Children may*	*Children may*
• Use obvious sensory information to explore materials (e.g., feel a pine cone; look at a rock; listen to a bell). • Watch or do something but not react in any way other than looking or repeating an action (e.g., watch the snow fall; repeatedly throw a stone in the water).	• Use multiple senses to explore materials (e.g., examine a feather by blowing on it, stroking it on their cheek, and smelling it); begin to show an interest in processes (e.g., observe how a plant grows). • Observe how something works or when something happens and indicate or comment on what they see (e.g., after throwing a stone in the water, say "Look. There are circles"; "My car went faster that time").	• Use multiple senses to explore the natural and physical world in greater detail (e.g., notice the soil is different colors, has large and small stones, smells damp, has little bugs crawling in it, and is warmer in some areas than in others). • Use observation to understand how tools and materials work (e.g., after throwing a stone in the water, say "When I threw the rock, it made circles on the water"; "When I put the block under this end, the car went down the ramp faster").
To support children's current level, adults can	*To support children's current level, adults can*	*To support children's current level, adults can*
• Provide a variety of simple sensory materials (sand, water, shaving cream, spice jars, sandpaper, fingerpaint, percussion instruments). • Watch alongside or copy children's actions (e.g., throw stones in the water together with a child).	• Emphasize multiple sensory aspects of materials (e.g., color, texture, and smell of paint) and processes (e.g., changes in color, texture, taste, and smell while cooking). • Encourage children to describe the actions and processes they observe in their own words (e.g., "What's happening with the soap?"; "What happened when you added one more block at the top of the ramp").	• Provide opportunities for children to use multiple senses while exploring the natural world (e.g., walk in the woods, dig in a garden) and physical world (e.g., observe the appearance and temperature of melting icicles or puddles drying in the sun). • Encourage children to describe what happens when they or you act upon objects ("I wonder what we might hear if we take a deep breath before blowing into the opening"); ask them to tell you what to look for ("What will I see when you lift up the other end?").
To offer a gentle extension, adults can	*To offer a gentle extension, adults can*	*To offer a gentle extension, adults can*
• Talk about clearly observable sensory properties such as appearance (color, size, shape), sound (loud, soft), or texture (rough, smooth, scratchy, sticky). • Describe what you observe (e.g., "Wow! When I throw the stone in the water it splashes and makes circles").	• Hunt for things with similar sensory properties (e.g., things of the same color or with similar sounds) and contrasting sensory properties (e.g., things that are shiny or dull; things that are loud and soft). • Provide materials children can transform to observe and describe the changes (e.g., seeds, cornstarch and water, play dough ingredients); provide new vocabulary words for their observations (e.g., *ripple, liquid, solid, germinate*).	• Encourage children to describe in greater detail what they are observing with their senses (e.g., "What else do you notice about the cheese crackers?"). • Ask children why they think something happened or to speculate what would happen "if" (e.g., "The water tastes sweet. I wonder what would happen if we added salt").

CHAPTER 4

KDI 46. Classifying

G. Science and Technology

46. Classifying: Children classify materials, actions, people, and events.

Description: Children group similar things together. They identify relationships between things and the categories they belong to. Children look for new ways to organize the knowledge they already have and for ways to fit new discoveries into familiar categories.

At large-group time, Kacey sorts through the basket of streamers until she finds a green one that matches her shirt.

At work time in the book area, Tasha puts all the black checkers on the black squares and all the red checkers on the red squares.

At work time in the toy area, Tracy fills a pegboard with red pegs on one half and yellow ones on the other. "I made a birthday cake," he tells his teacher Linda. "Blow out the candles!"

At work time in the toy area, Douglas makes a pile of all the metal things he has gathered around the room "for the magnet to stick to."

At outside time, Dierdra wonders why some flowers have smells and some don't.

At small-group time, Josh sorts his dinosaurs into two piles. He explains to Emily (his teacher), "Those are plant eaters, and those are meat eaters."

Children, like adults, classify things to help organize their lives. Sorting often occurs during play because it is a means to an end. For example, children may sort out all the small cars to roll on a track, or pick out all the red beads to string on a necklace. With the vast amount of information preschoolers process each day, categories also help them store and retrieve the knowledge they are accumulating. The more they learn, the larger and more diverse these categories become. As children continue to make discoveries and adjust their thinking about the properties and functions of objects, events, and people, their categories and what goes into them also change over time. For example, the large category "animals" becomes divided into subcategories such as farm animals, zoo animals, and pets.

In learning about the properties of things and sorting them into groups, young children use all their senses. Preschoolers sort things into common categories (red and blue, big and little, hard and soft) based on the characteristics they observe and the labels others provide. In addition, they develop their own systems of classification and sort information according to the categories they create, some of them quite imaginative (fuzzy and bumpy, jingly and bong-bong sounds). Research also shows that young children develop "intuitive theories" to help them decide how to classify things (Gelman, 1999). For example, if they sense that appearance is important (as it might be with doll clothes or small figures), they will classify objects on the basis of a visual attribute such as size or color. If, on the other hand, they sense something's function is what matters (as might be the case with tools), they will sort on the basis of how something works or the way it is used.

How Classifying Develops

The developmental sequence for classifying can be observed in children's spontaneous behaviors (Langer, Rivera, Schlesinger, & Wakeley, 2003). They begin by separating objects from a pile or grouping things together because they share an attribute, for example, selecting all the red beads. They may or may not be able to state the reason for the grouping and do not always apply the rule consistently. For example, if they begin with small and large red beads, they may switch midway to large beads.

At small-group time, Jonah fills a cup with paper and cloth pieces and tapes it closed. He shakes the cup and says, "It doesn't make a sound. It has only quiet things in it."

❖

At snacktime, Gigi says to Peter, "You and Sam both have glasses."

❖

At work time in the house area, Samantha looks for the "softest" blanket to wrap her baby doll.

At the next level, children sort consistently and use the words *same* and *different* to match and compare attributes. As they apply concepts acquired in mathematics to science, they also understand and use the words *some, none,* and *all* to describe properties and collections.

At small-group time, Jayla sorts her dinosaurs by color, picks up two yellow ones, and says, "They are both yellow, but this one has horns and this one doesn't."

At work time in the toy area, Noah separates the red from the blue snap-together blocks, then builds a house with the blue blocks and a garage with the red ones.

At work time in the block area, Celia builds two towers and tells her teacher, "They are the same."

At work time in the house area, Julia is playing doctor. Looking over the patients in her waiting room, she tells them, "If you are sick, you sit here. If you are all better, you sit there."

At recall time, Che talks about a Lincoln Log house he built with Joshua: "I didn't put on all the pieces. Joshua put on some."

Children at this developmental level can also identify things that do not belong to a set; for example, they divide beads into those that are "red" and "not red." Children need to hear the word *not* many times to grasp this fundamental distinction, another example of how language and thinking are interconnected. Understanding the concept of "not" is important in mathematics as well as in science, particularly when it comes to collecting and making sense of data. Mathematics researcher Juanita Copley (2010) explains, "This two-part type of classification — *has* versus *has not* — is fundamental in collecting certain kinds of data" (p. 143). Being able to classify things in this way helps young children understand basic mathematical and scientific principles, and allows them to use simple graphs and charts to represent the information they collect.

At work time at the sand table, Corey says, "I don't want to play with those cups. They don't have handles."

❖

At cleanup time, Dasha finds a wooden inch-cube among the wooden beads and says, "It doesn't have a hole. It doesn't belong here."

❖

At work time in the house area, Nicole and Miko are baking a cake. Miko says, "We're only putting in pine cones and yellow sponges. We don't want any of those white things [packing peanuts]."

Next, children are able to sort a collection of objects based on more than one attribute, such as color and size. For example, they can sort shapes into "red" and "not red," and then further sort each color group into small and large shapes. Their reasons for grouping objects may not always be obvious to someone else, so adults should ask rather than assume that they

Encouraging children to state the attributes behind their categories helps them clarify their thinking and become more confident in their own reasoning abilities.

know the basis for children's classifications. In addition, children's groupings may be somewhat unstable at first. If they see another child sorting on a different basis or a third attribute is introduced (a triangle is added to a set of shapes that is otherwise all rectangles), they may be confused about how to proceed. Encouraging children to state the attribute(s) behind their categories helps them clarify their thinking and become more confident in their own reasoning abilities. For example, you might say, "It looks like you have all your big brown horses in this barn and the rest of the big horses over there, or did you have something else in mind?"

At snacktime, during a conversation about pets, Drake tells Susannah, "Puppies are little dogs."

At work time in the toy area, Hannah finishes a puzzle and explains to her teacher, "I knew that piece fit because it had the same color on it and the same shape."

❖

At small-group time, Farrah makes two piles of colored magnets. Pointing to one pile, she says, "These are all the biggest shapes and they're all blue." Pointing to the other pile, she says, "These are all small shapes but they are different colors."

The highest level of sorting is when children can recognize and describe the reason behind the sorting even when someone else has done it. In this instance, children are not actively handling the materials or choosing the rules. They have to perceive one or more attributes that the groups share in common ("These are all small red beads") and also determine that the attribute(s) are not shared by objects in the other group ("None of these are small red beads"). Children at this level can also describe whether and what type of object can, and cannot, be added to each group.

Teaching Strategies That Support Classifying

Young children love to explore the similarities and differences between objects. They enjoy making collections, sorting things into categories, and comparing their attributes. To promote and enhance children's spontaneous interest in classification, use the following strategies.

Encourage children to collect and sort things

Assembling, sorting, and re-sorting collections are among young children's favorite pastimes. They create and classify collections based on sensory properties (plain and scented markers), how they use them in play (magnets too weak or strong enough to stick two items together), likes and dislikes (strawberry cream cheese is good but the kind with chives isn't), and relationships to themselves (everyone in my small group is invited to my birthday party). There are many opportunities during the day and throughout the indoor and outdoor learning environment when preschoolers can indulge their passion for collecting and classifying objects and experiences.

At small-group time, Len points to the longest Cuisenaire rods and says, "These are the dinosaur bones." Holding one of the shorter ones, he says, "These are the monkey bones."

At work time in the toy area, Rene points to the two stacks of puzzles she's made and explains to her teacher, "These are the easy ones I can do, these are the ones that are too hard."

At outside time, Trent observes, "Some of the balls bounce and some don't."

Most important, an abundance of diverse materials in the classroom lends itself to classification. By labeling areas and materials, you provide children with an initial basis for categorizing things. Labels help them think categorically during the plan-do-review sequence (playing with "puzzles" in the "toy" area), including knowing where to find and return things (books are on a shelf in the book area, except for the really big books which are in a milk crate; in the art area, paint brushes are in a container but markers are stored in a basket; blue tops go on the blue markers). Diverse materials, especially those that appeal to a variety of children's senses, also allow them to create their own classification systems and labels (e.g., "slidy" versus "rolly" toys).

At the end of small-group time, Rosie sorts her materials into the three tubs her teacher left on the table: one holds glue sticks, one contains the collage materials from outside, and one is for paper.

The idea of classifying or sorting materials can also be emphasized when you introduce new items to the classroom. You might do this at small-group time or during message board time, and then let children know in what area and with what other materials they can find the new objects. Share one or two physical or functional features of the materials that explain why they will be placed in these locations. Ask the children why they think items do (or do not) belong in a particular area or grouped with other materials. Once children are familiar with the classroom, you can ask them where a new item should be stored, and why they think it belongs there and not somewhere else.

At the end of a math small-group time, the teacher asks the children where to store the dot cards for those who want to use them at work time. Some say the toy area because children can put beads on them. Others suggest the art area because children might want to color them. In the end, they put them in the toy area because they're like playing cards.

In addition to creating collections in the classroom, children can also make collections at outside time. Nature provides many materials for them to sort — fallen leaves, seeds, twigs, shells, and pebbles. Children can also classify the things they see, hear, and smell (things that fly or crawl, trees with leaves or needles, flowers that smell and those that don't). Walks around the neighborhood and field trips provide additional opportunities to make interesting collections. Give each child an easily carried container (a small bag or bucket) to collect items as they walk. Back in the classroom (at small-group time the next day), encourage children to sort and label their collections. Then make the items available for children to use at work time in whatever location(s) is appropriate to their attributes. For example, acorns can be placed in the house area for pretend cooking, stored in the art area for gluing on paper, or hidden in the sand table.

Call attention to "same" and "different"

Plan activities and take advantage of spontaneous occasions during the day to focus children's attention on materials and experiences that are the same and different. To follow preschoolers' developmental progression in classification, start out with sets of objects that are different (counting bears and small cars). Then focus on a single attribute that is the same (all the red buttons), followed by objects that differ by only one attribute (red buttons and blue buttons). Later

Teachers can draw children's attention to materials and experiences that are the same *and* different, *focusing on one, two, or several attributes depending on the child's developmental level.*

you can help the children work with two attributes (large red buttons and small blue buttons). Lastly, call their attention to things that are the same in some way but different in one or more other ways (all the buttons are red, but some have two holes and some have four holes). With these materials, you can also introduce the concepts of *some, none,* and *all.*

At work time in the block area, Penny matches wooden animals (bears with bears, monkeys with monkeys) and says, "They are brothers. They are the same."

❖

At work time in the toy area, Evan holds up white and silver toy airplanes and says, "These are the same but they are different colors."

❖

At work time in the house area, Carlos and Kyra look for more pink "jewels" to decorate their cupcakes. Kyra looks in the refrigerator and says, "None here. Any in the cupboard?"

❖

At work time in the block area, Amanda points out a block that she calls a "special chair." "It's square and it's blue," she tells her teacher. Her teacher repeats, "It's square and it's blue." Amanda nods. The teacher holds up a red square block and Amanda says, "That's a red square. It's not special."

❖

At outside time, Andy pours sand into a plastic jar using a funnel. When he finishes, he says, "It didn't all get in. Some spilled here, see?"

As you create a daily plan, think of opportunities to introduce the idea of same and different. For example, after reading a favorite book at small-group time, provide art materials for children to draw characters that are the same and/or different from those in the story ("I wonder if your monsters will look like the ones Max saw, or if they will look different"). At work time, play games such as concentration and bingo to help children match shapes, numbers, and letters. During cleanup time, play I spy ("I spy something in the toy area that's the same shape as this [hold up a circle] and needs to be put away"). Go on treasure hunts in the classroom, playground, or neighborhood to look for objects that are the same and/or different in one or more ways from a comparison item (things that are red, doors made of wood with glass panes, leaves that are green or yellow, street signs with letters and numbers). Challenge children to imitate the same movement as another child and to come up with a different movement when it is their turn to be leader. Use the concepts *same* and *different* as the basis for transitions ("Everyone with gloves go outside. Now everyone with mittens"). Play guessing games featuring same and different attributes ("I see something in the house area that is the same color as Mary's shirt. What do you think it is?" or "I see something different about what Curtis and Shamar have on. I wonder what you see that's different about their clothes").

At work time in the art area, Joey draws two monsters on his dry-erase board. "I made two monsters," he tells his teacher Sam. "This one has sharp teeth and this one doesn't." Sam points to the monster with the zigzag mouth and says, "They're the same because they are both monsters, but also different because this one has sharp teeth." Joey nods.

Spontaneous conversations also provide openings to explore same and different. At snacktime, serve a variety of foods and encourage children to identify those that look, taste, or feel the same and to compare them to those that produce different sensations. When children bring up their families, talk about similarities and differences, such as the number of people and who they are, whether or not they have pets, what holidays they celebrate, and the kinds of music or food they enjoy at home.

To help children think further about and describe these concepts, ask them to provide instructions on how you can do or make something the same as or different from theirs. For example, as you work alongside children, you might say, "How can I build a tower the same as yours?" or "I want to use different color pegs on my pegboard. Which colors should I use?" Make an occasional mistake to encourage children to identify the characteristics of the object or action that will make yours the same or different. For example, if you are imitating a movement in which the child puts a foot in front, move yours to the back. If the child does not say anything, you might comment, "I don't think that looks the same. Help me figure out what I'm doing different."

Use "no" and "not" language

As described earlier, classification also involves recognizing when something does not possess a particular attribute (families that have pets and families that do not have pets; trail mix with raisins and trail mix with no raisins). To develop the concept of *no* or *not*, children need to hear those words and have experiences that illustrate them many times. Fortunately, there are numerous opportunities to highlight this idea as children work with materials and engage in various actions throughout the day. For example, when children dress to go outside, you can ask them to identify whose jacket has a hood and whose does not. At snacktime, comment on who does and does not want juice. When children make collections, encourage them to identify the attributes that items in a group do not have ("These beads are shiny. Those are not shiny").

At small-group time, Jonah finds a small pterodactyl dinosaur in each color. When Sue (his teacher) hands him a green T. rex dinosaur, Jonah says he only wants pterodactyls.

At snacktime, Jeff holds up two halves of the apple he has just cut. "Look. This part has seeds and this part has no seeds."

At greeting time, after reading the book If You Give a Pig a Pancake, *Todd says to his teacher, "I like waffles, not pancakes."*

At outside time, Martin makes a pile of small stones. He dips each one in a pail of water and examines it carefully. He puts the stones that change color when they are wet into one bucket and the stones that do not change color when they are wet into another bucket.

You can also introduce children to the universal "no" symbol (a red circle with a diagonal line superimposed on an image of the object or action at issue). An example is the "work-in-progress" sign that children can choose to put on an art project they want to continue working on the next day. The image of a hand with a line through it signifies "Do not touch" and sets the object apart from that category of items that are available to be handled.

Here are some other ideas for familiarizing children with the concepts of *no* or *not:*

- Children who are or are not in school that day
- Weekends and holidays when there is no school
- Areas children plan or do not plan to work in
- Materials children did or did not use at work time
- An area of the classroom where something does not belong
- Songs with and without fingerplays
- Foods that children do and do not like
- Names that do or do not begin with a particular letter
- Words that do or do not rhyme with one another
- Materials children have played with or never played with before
- Things that do or do not make noise
- Movements that do or do not require use of the feet
- Days when it is raining or not raining

For examples of how children classify materials and events at different stages of development, and the strategies you can use to scaffold their sorting behavior, see "Ideas for Scaffolding KDI 46. Classifying" on page 42. The chart offers additional ideas on how you might carry out the strategies described in this chapter during play and other interactions with children.

Ideas for Scaffolding KDI 46. Classifying

Always support children at their current level and occasionally offer a gentle extension.

Earlier	Middle	Later
Children may	*Children may*	*Children may*
• Match one object to an identical object (e.g., find a cup that is identical to another cup); group similar things together (e.g., take a block out of the dinosaur basket and put it in the block basket). • Identify things as being the same or different when they match or sort (e.g., "These cups are the same"; when removing a block from the dinosaur basket, say, "That doesn't go there").	• Sort based on one attribute (e.g., select all the red blocks from a pile; divide a group of shells into large ones and small ones); provide a basis (reason) for sorting (e.g., "Straight ones go here and curly ones go there"). • Use the words *some, none,* and *all* when they sort (e.g., "These are all the mommy's babies but only some of them are girls"; "There are no purple beads in the basket").	• Sort based on two or more attributes (e.g., choose buttons that have four holes and are also shiny from a button jar with many types of buttons). • Use the word *not* to identify a property something does not have when they sort (e.g., "Ten cars have numbers on them and two don't").
To support children's current level, adults can	*To support children's current level, adults can*	*To support children's current level, adults can*
• Provide sets of materials so children can find matches and group similar items together (e.g., animal and people figures, tableware). • Affirm when children identify things as being the same or different (e.g., "Yes, they're the same because they're both cups"; "A dinosaur *is* different from a block").	• Describe and encourage children to describe the attribute on which they sort (e.g., "I wonder why you put these bears in this basket and those bears in the other basket"); ask children how to make a collection just like theirs. • Include the words *none, some,* and *all* in your conversations with children (e.g., "Some of us have red hair"; "All of us are running with our arms in the air!").	• Provide objects that vary along two or more dimensions (e.g., blocks in several sizes, shapes, and colors); ask what else is similar about the objects children group together. • Emphasize the concept *not* in your conversations with children (e.g., "Five children are going to kindergarten next year and three are *not* going. They're not old enough").
To offer a gentle extension, adults can	*To offer a gentle extension, adults can*	*To offer a gentle extension, adults can*
• Match objects and create groups of similar items; describe what they are doing (e.g., "I'm putting all the spoons in one pile"). • Ask children what else is the *same* or *different* about materials (e.g., "What else is the same about these counting bears?") and how you or they can do the same or a different action (e.g., "How can we move our arms differently than we just did?").	• Sort objects based on two attributes and describe what they are doing (e.g., "I'm picking out all the orange pegs that also have holes on top"). • Ask children to give you *none, some,* or *all* of something (e.g., "Please hand me all the pink scissors"; "I need the small shoes but none that have laces"; "Put in some but not all of the water").	• Provide opportunities for children to attend to two or more attributes (e.g., at transitions, say, "If you're wearing sandals and a barrette in your hair, go to the rug"). • Ask children what else is *not* the same about objects or actions (e.g., "What else is not the same about Ben's jacket and Ilana's jacket?").

CHAPTER 5

KDI 47. Experimenting

G. Science and Technology
47. Experimenting: Children experiment to test their ideas.

Description: Children experiment to test whether an idea is true or a solution will work. They may encounter problems with materials that they do not have answers for. They experiment by manipulating materials, using trial and error, and then approaching the problem with possible solutions in mind.

At outside time, Amelia wonders what would happen if she used colored water on the plants. She adds red food coloring to her bucket and pours it on the lettuce. "Still green," she observes. She puts other colors in the bucket and carefully studies the lettuce leaves each time she waters them.

At work time in the toy area, Dov shows Sam the "wrist thing" he's made with pipe cleaners and explains, "It's waterproof. It can stay in the water a long time." They decide to immerse it in a bowl of water to check it out. After counting up to 20, Dov fishes it out, refastens it on his wrist, and says to Sam, "See. It still works."

At work time in the block area, Yasmina is building a house out of small unit blocks. She wants a roof "to keep out the rain," but when she rests a block on one wall and lets go, the roof block falls into the house. She tries to balance the roof block on the opposite wall, but again it falls in. Karl, who is building nearby, offers Yasmina one of his longer blocks, but she does not take it. Instead, she moves the walls of her house closer together. Then she sets the roof block across them. It is still a bit too short so she moves the walls even closer, and this time the roof fits just right!

Young children experiment for two basic reasons. They investigate materials and processes out of curiosity. For example, they see something (a leaf) and wonder what will happen if they act on it in some way (blow on it, tear it, or crush it between their fingers). Children also experiment to solve the problems they encounter in play. They try one or more ideas as they look for a solution that works (will a block of a different shape or size balance? will the puzzle piece fit if they turn it?). In both types of experimentation, preschoolers are actively engaged in handling materials, observing the outcomes of their actions, and forming ideas about the world. Like adult scientists conducting experiments on topics that interest them, preschoolers also "learn by doing":

As any scientist knows, the best way to learn science is to do science. This is the only way to get to the real business of asking questions, conducting investigations, collecting data, and looking for answers. With young children, this strategy can best be accomplished by examining natural phenomena that can be studied over time. Children need to have a chance to ask and answer questions, do investigations, and learn to apply problem-solving skills. Active, hands-on, student-centered inquiry is at the core of good science education. (Lind, 1999, Introduction)

How Experimenting Develops

In the process of experimenting, children observe cause-and-effect relationships. At first this happens accidentally, later with conscious awareness and intent. Thus infants and toddlers initially experience an effect as a random event and try different things to repeat it. They may eventually be successful in these efforts. For example, while waving their arms, babies may accidentally hit a mobile suspended above them, which in turn makes the shapes jiggle. After this happens several times by accident, they may start to wave their arms more. They do not yet understand that it is their hands hitting the mobile that causes the motion, but they have made a connection between arm waving and jiggling shapes. By contrast, three- and four-year-olds are increasingly aware of causality, that is, the idea that a specific action results in a predictable outcome. Rather than leaving such actions to chance, they seek to establish the nature of the link so they can intentionally perform the causal behavior again.

During work time at the sand and water table, Nolan says, "When you add water to the snow, it melts!" He repeats this action in each corner of the table.

At small-group time, Kovid moves a feather across the floor by blowing air through a straw. He says, "I made it roll over." Then he sticks the feather into the end of the straw and says, "It will pop out." He blows to show how the feather pops out of the straw.

At recall time, Avram reports that he played with soap suds and a straw at the water table: "When I blew fast it made little bubbles, and when I blew slow it made big bubbles."

At small-group time, Erin grinds apples in a food mill for applesauce. She looks underneath and sees that nothing is coming out. "You have to grind it harder," says Mike, demonstrating with his grinder." "I think it's stuck in the holes," says Erin. Sandra calls from across the table, "You gotta scoop out the peel." Erin tries this and the ground apples flow freely again.

Preschoolers' ability to mentally represent something in both its "before" and "after" state helps them make this bridge between a cause and its effect. They recognize that what they or someone (or something) else did in the middle — between the before and after condition — is what resulted in the change. Using their growing store of experience, preschool children form ideas about what and how an effect occurred. They can then test out this idea — experiment — to see if their theory is true. Based on their emerging ability to picture or imagine the possible outcome of an action, they can also begin to predict the outcome of an action that has not yet occurred (see KDI 48. Predicting).

At work time in the house area, the "mommy" and "daddy" need to take their sick baby doll to the hospital. The daddy gets a wagon. "It will be too bouncy," says the mommy. "Let's carry the baby." She wraps the doll in a blanket and hands it to the daddy, saying. "The baby is sore. You have to carry it gently." The daddy walks slowly and confirms that "The baby isn't bouncing."

❖

At work time in the block area, as Jeremy makes a tower of blocks with Ben, he says, "It's wobbling. It will fall down if we put more on top."

Children's use of trial and error in experimenting. While preschoolers are increasingly purposeful in these early experiments, they still proceed largely by trial and error rather than systematically. Even if children persevere, they may not try various options in logical order. For example, if one spoonful of paint does not produce the color they want, they may add three more spoonfuls instead of just one more. Nevertheless, a growing capacity for observing the properties of materials, and the relationships among them, helps young children become increasingly thoughtful about the nature and order of the solutions they try.

At large-group time, Eddie tries a variety of scarves and streamers until he finds the one that billows just the right amount as he twirls around.

❖

At small-group time, Vye takes the "shaker" (small-holed) lid off a glitter container and holds it over her paper. When the glitter pours out, she says, "Ooh. Too much." As she scoops some back inside with a spoon, she sees Conan successfully shaking his lidded container vigorously over his paper. Vye rescrews her shaker lid and tentatively shakes it over her paper. When she observes that the result is what she wants too, she continues to shake out glitter.

❖

At outside time, Lon tries to drive his truck through the sand but the wheels get stuck. "You have to pack it [the sand] down," explains Myrna, and shows him how to do it. Lon gives two pats but his truck still gets bogged down. "I think I have to pat it more," he says and this time the truck rolls smoothly. "Pack it really hard," he tells Corey, who has just joined them in the sandbox.

❖

At work time in the toy area, Bonita dumps the 10-piece nesting puzzle on the tabletop. After reassembling a few boxes in the right order, she sees she's left a gap (between the 10th and 7th piece). She takes apart what she's done so far and spreads out all the pieces. Then she slowly arranges them in size place, sometimes instantly

As they experiment, children begin to notice changes that occur in the natural and physical world, including the permanence or duration of an effect, such as how the clay becomes soft when you add water, then hardens with the passage of time.

seeing where one fits in the sequence and other times comparing them side by side. When they are all lined up, she starts with the biggest box and quickly reassembles all 10 pieces.

Children's observations of changes over time. Preschoolers are increasingly aware that the passage of time plays a role in the natural and physical world (Van Scoy & Fairchild, 1993). For example, preschoolers pay attention to sequence when they experiment. They notice that the cause comes first and the effect follows (after you move the magnet closer, then the screws are drawn to it). Children also observe the permanence or duration of an effect, that is, whether it lasts, fades away, or changes to yet another state (the water stays slippery after you add dish detergent, but the bubbles dissolve unless you blow new ones). Finally, children notice how quickly something happens, that is, the pacing or speed of events (e.g., puddles fill quickly in a heavy rainstorm; snow melts slowly into a puddle). Older preschoolers, who can hold images in mind for a longer time, might even notice delayed effects ("When I dripped water on the paper, only the area underneath got wet; when I came back later, the wet area had spread all the way to the edges").

At recall time, Arvin describes how he made his picture. "First I put glue on the beans, but they didn't stick too good. Then I put glue on the paper and dropped the beans on top. The second way works lots better."

❖

At snacktime, Linette describes her family's weekend camping trip. "I got stung by a bee and I cried because it hurt. My dad put baking soda on it and it didn't hurt anymore. Baking soda is good medicine."

❖

At greeting circle, Ella and Faye are reading a book. "If we skip some pages," says Ella, "we'll finish faster."

❖

During outside time, Shawn says to Farrar, "Let's get a bunch of acorns and hide them under the bush. Then tomorrow we'll see if the squirrels ate them!"

As children's measuring skills (KDI 36. Measuring) in mathematics develop, they begin to observe the amount that things change over time. In their scientific investigations, they explore how actions related to sequence, duration, and pacing change many other properties of the objects that interest them.

At daily team planning, one teacher tells her coworker: "Ricky likes playing with things he can change to see an immediate effect, like mixing paints and screw-together toys. He might also be interested in sponges and water, a metronome, flashlights, anything in squeeze bottles, windup clocks, and balloons." The teachers agree to try one or two of these items at small-group time and then add them to the interest areas.

❖

At work time in the art area, Raisa tries to attach a feather to a shoelace. She applies glue to the feather but it falls off, so she squeezes some on the shoelace. There is now lots of glue on the table, but when Raisa lifts the shoelace, the feather still does not stick. When the teacher asks her what else she could use, Raisa looks on the art shelf and sees tape. She says, "Tape is sticky." A look of satisfaction appears on her face when she lifts the shoelace and the feather, now securely attached with tape, dangles from the end.

Teaching Strategies That Support Experimenting

Young children are naturally inquisitive and show a spontaneous interest in figuring out how things work. Conducting experiments is one way in which they ask and answer questions about the world and solve problems that arise in the course of their play. To help preschoolers carry out meaningful experiments and learn from their investigations, use the following teaching strategies.

Ask and answer *"What if...?" "Why...?"* and *"How...?"* questions

Science invites the use of comments and questions such as "I wonder..." and "What do you think would happen if...?" (Neill, 2008). Children are encouraged to experiment and develop higher order thinking skills when you observe alongside them and follow up their attempts with open-ended "How...?" and "Why...?" questions. For example, you might ask questions such as "How did you get the top to spin?" "Why do you think it kept falling?" "What else could

Adults can scaffold children's learning by talking with them about what they are doing and discovering as the result of their actions.

you do to make it stick?" "How can we get the sled to go faster? How can we make it slow down?" "Why do you suppose this one worked but that one didn't?"

In addition to asking questions yourself, make comments that will prompt the children to ask questions of you and of themselves. By encouraging and welcoming their inquiries, you support and stimulate children's natural curiosity (Gronlund, 2006). If you express interest and surprise ("Wow! It bubbled over the top when we added vinegar to the baking soda!"), the children will be inspired to ask on their own why or how something happened. You can then ask follow-up questions to help them experiment, test out their ideas, and try to answer their own questions.

A group of children and their teacher are rolling and cutting play dough to "bake cookies." The children observe that the green dough is dried out and crumbles, but the red dough rolls out easily and doesn't fall apart when they use the cookie cutter. The teacher wonders why this is so. The children offer the following explanations: "The red play dough is older" (made before the green) and "The green lid was not put on and it dried out." The children like this last idea and are ready to accept it, but the teacher wonders how they could find out for sure. They decide to leave some play dough out overnight and return in the morning eager to feel it and confirm whether their hypothesis is true. It is!

The matter could end there, but one child asks why leaving the lid off makes the play dough dry out. Where does the water go? Another child says that when his mother adds water to dried-out play dough at home, it gets soft again. The children try this with their play dough and observe that it gets softer but is now "too squishy."

Encourage children to gradually replace trial-and-error exploration with systematic experimentation

An abundance of diverse materials and the freedom to explore them guarantees that young children will experiment in the classroom and outdoors. They will learn about materials and their properties, and how they act upon one another. However, this information may be superficial and disorganized unless adults act with intention to help them reflect on their observations and gradually construct more lasting ideas. Young children rely on adults to help them move beyond random and chance discoveries to more systematic investigations and critical thinking.

To scaffold learning during experimentation, talk with children about what they are doing and the results of their actions. Describe and encourage them to describe materials and outcomes. Use small-group times to carry out simple hands-on experiments with easily observed results (mix different amounts of flour and water and compare the consistency; paint with water outdoors and watch the pavement dry; listen to the sounds made by shakers the children fill with birdseed, sand, and gravel).

In addition to introducing novel materials and tools, encourage children to consider alternative ways of using and combining familiar materials. Ask what is different about what they did this time and the effect(s) their actions produced. Offer challenges that suggest step-by-step or systematic changes in their investigation ("Could you make it go just a little bit faster?" "How about a medium amount faster?" "What if you wanted it to go really fast?").

At snacktime, Suki and Rachel set the table. They want to give each person two plates (one for red apple slices and one for green apple slices) and a bowl (for yogurt dip). After laying out one place setting, Suki looks worried. "What if there's not enough room at the table?" she asks Rachel. Their teacher asks how they can find out. The girls decide to begin with one plate and one bowl. It just fits. "What if children want to separate red and green apple slices?" asks the teacher. The girls decide their classmates can make two piles on either side of their plate.

After each step in their investigations, help children recall the "before" situation, their actions, and the "after" condition. In addition to verbalizing children's observations, adults can help children represent the experimental process and results through artwork, journal dictation, or photographic records. Provide books and other appropriate resources they can use to answer their own questions.

At work time in the toy area, Miguel puts a block on one side of the balance scale and three beads on the other. When he tries to add more beads, they roll off. "I wonder if you can put the beads on the scale a different way," says his teacher. Miguel decides to set a plastic cup on the scale and put the beads in one at a time until it balances. Then he takes the cup off the scale, counts the beads, and asks his teacher to write that it takes 16 beads to weigh the block.

Provide materials and experiences for investigating how things change with time

Time can be mathematical — how long does it take for something to happen? But time can also be scientific — how do the properties of things change over time? Many aspects of the natural and physical world undergo change with time — kittens grow bigger, leaves turn brown, sand compacts as it's walked on day after day. Adults can provide materials and experiences that help draw children's attention to the role that time plays in these transformations.

Since time can be abstract, preschoolers need to make it concrete by using time-related materials. They experiment with time by playing with toys and tools that can signal stopping and starting (timers, stop signs, musical instruments), and things they can set in motion (wheeled toys, metronomes, balls, tops, pinwheels). Each movement is a concrete representation of time. For example, if they turn the dial on the timer a little bit, it rings sooner than if they twist it a lot. How fast and long the pinwheel turns depends on how hard they blow or how strong the wind is.

Encourage children to notice how long it takes for something to happen. "I put on my coat, boots, and mittens before you got your coat on." "Yesterday we played the cleanup music two times before we were done. How many times do you think we will have to play it today?" As you talk about "before" and "after," encourage children to observe when something does not change as well as when it does. "I wonder if it will look different after you stir it or if it will look the same."

Finally, help children experience and experiment with the role time plays in nature. For example, they might investigate how shape — round and flat pebbles — affects the speed with which objects roll downhill. Or they might notice how quickly items of different weight (acorns and feathers) fall from various heights. You can plant a garden and talk with children about the order in which seeds germinate and how the plants change in appearance over time. Call their attention to how quickly snow melts or pavement dries in open, sunny areas versus protected, shady spots. Ask open-ended questions that encourage children to consider how they can test their theories about time:

KDI 47. Experimenting 51

Encourage children to experience and experiment with the role that time plays in nature. Here children examine a sunflower at the end of its growing cycle, exploring the seeds it has produced.

A day after it snows, the children notice that some areas of the playground are bare and others still have mounds of snow. "I wonder why that is?" asks their teacher. A few children suggest that the sun makes snow melt faster. "How can we find out?" the teacher asks. They decide to place snowballs in two locations and check them later. At pickup time, they see that the snowballs in the shady area remain unmelted while the snowballs in the sunny spot have disappeared.

For examples of how young children investigate scientific phenomena, and how adults can support and gradually extend their learning at different stages of development, see "Ideas for Scaffolding KDI 47. Experimenting" on page 52. Use the additional ideas in this chart to help you carry out the strategies described in this chapter in your play and other interactions with children.

Ideas for Scaffolding KDI 47. Experimenting

Always support children at their current level and occasionally offer a gentle extension.

Earlier	Middle	Later
Children may	*Children may*	*Children may*
• Accept what happens when they work with materials without questioning how, what, or why. • Use materials to find out what they can do and to see what happens with them (e.g., pour water through a water wheel).	• Ask simple questions about what happens when they work with materials (e.g., When a block tower falls down, ask "What happened?"). • Use trial-and-error to investigate materials (e.g., pour different amounts of water in the water wheel and notice that sometimes the wheel turns faster than other times).	• Ask questions about what happens when they work with materials and experiment to find an answer (e.g., wonder why their block tower fell down and experiment to find different ways to build it). • Express an idea and experiment to test it out (e.g., Say "I think if I spin the wheel, the water will go through faster" and then try it out to see if it works that way).
To support children's current level, adults can	*To support children's current level, adults can*	*To support children's current level, adults can*
• Explore alongside children and imitate their actions with materials (e.g., mix paints in similar proportions). • Provide interesting materials for children to manipulate and see the effects.	• Ask children to describe what happened (e.g., "Tell me what you just saw"). • Encourage children to describe the different materials and actions they tried in their investigations (e.g., "Tell me the different things you tried to get the tent to stay up").	• Encourage children to answer their own questions by experimenting with materials (e.g., "What could you do to find out what made the block tower fall down?"). • Acknowledge children's ideas and the fact that they are experimenting to test them out (e.g., "You're going to experiment by spinning the wheel in different ways to see if the water goes through faster").
To offer a gentle extension, adults can	*To offer a gentle extension, adults can*	*To offer a gentle extension, adults can*
• Call attention to what happens with materials as you work with children; wonder what happened (e.g., "I mixed red and white and got pink"; "The sand castle caved in. I wonder what happened"). • Comment on what children do with materials and what happens (e.g., "You're pushing the stapler up and down"; "When you poured water through the top of the water wheel, the wheel turned").	• When children wonder what happened, encourage them to recall what they just did (e.g., "What did you do just before it fell down?"). • Wonder what other ways children can think of to investigate materials (e.g., "I wonder what else you could do to figure out how to get the tent to stay up").	• Refer children to one another to answer their questions (e.g. "Sam's other sand castle also collapsed, but the one he's building now is staying up. I wonder what he did differently"). • Ask children about the results of their experiments (e.g., "What happened? Did your idea about spinning the water wheel work?").

CHAPTER 6

KDI 48. Predicting

G. Science and Technology
48. Predicting: Children predict what they expect will happen.

Description: Children indicate through words and/or actions what they expect an outcome to be. They think about what happened in similar situations and anticipate what might happen.

At work time at the water table, Amani closes the lid on a plastic container filled with dishwashing liquid and says to Brian, "You shake it and it turns to bubbles." She demonstrates and says, "See? Bubbles, just like I said."

At work time in the art area, Wanda says, "I'm gonna cover this whole box, so I need lots of paint and a big paintbrush."

At outside time in the sandbox, Tamara says, "If you drive that truck through there [she points to a tunnel the children have made], it will collapse. It's too big."

At outside time, Nathan looks up at the gathering clouds and says, "We better go inside. It's gonna rain cats and dogs!"

For children, like adults, the more wide-ranging their experiences and the more familiar they are with a variety of materials, the easier it is to make predictions. However, even though a young child's knowledge is more limited than that of adults, preschoolers can still draw on their store of experiences to make an "educated guess" about the outcome of an action. They can also be relatively objective in their predictions, differentiating between what they *want* to happen ("If I add one more block to the high end of the ramp, my car will be the fastest") and what they *think* is likely to happen ("Even if I add another block, Tommy's car may still be faster than mine"). Prediction thus involves emotional maturity as well as intellectual capacity.

Children's scientific predictions are based on their observations (KDI 45. Observing) and experiments (KDI 47. Experimenting), and are thus closely tied to their developmental capacities in these other areas. Thus, if children's investigations are still based on trial and error, they are also more inclined to make hit-or-miss predictions. By contrast, preschoolers who are more focused in their observations and systematic in their experimenting are also more likely to be thoughtful and logical when making predictions.

At work time in the toy area, Hansen studies the animal picture cards face-up on the table. Before he turns over the next card, he says, "I think this will be a robin because I didn't get one yet."

❖

At small-group time, the children each mix a small bowl of batter for cupcakes. "It's too dry. It'll burn when I bake it," predicts Amy. The teacher asks what she can do to make the batter less dry. "Milk," says Amy. She adds a spoonful but that makes the batter too soupy. "More flour," decides Amy. Franco, working beside her, says, "My mom only adds a little at a time." Amy adds two spoonfuls of flour and Franco predicts, "I think one more is gonna do it!"

How Predicting Develops

The emerging cognitive abilities that allow children to make scientific predictions are similar to those they use when predicting the next event or outcome in a story (see KDI 21. Comprehension). They must recall relevant prior information (e.g., the rock sank to the bottom of the puddle but the Popsicle stick floated on top), reflect on its meaning or significance for what is happening in the moment (the twig is skinny and made of wood like the Popsicle stick), and then apply this information to form a mental image (representation) of what is likely to happen (the twig will float too).

At small-group time, Eli says, "I better not roll the play dough too skinny or it will break."

❖

At work time at the water table, while putting marbles down the water wheel, Jared says to Michelle, "That marble is too big. It gets stuck and nothing comes out."

❖

At work time in the block area, Josh stands up paper towel tubes like "bowling pins" and rolls a yarn ball towards them. He knocks over a few tubes, then says, "I need a different ball." He gets a larger rubber ball and tells his teacher, "This one will knock more down."

This is not to say that using these abilities makes children's predictions accurate, only that they are applying their current level of understanding to create a logical expectation. For example, a child using recall, reflection, and representation (a visual memory) may also observe that the twig floated on top of the puddle and reason as follows: A twig is from a tree. Other things from trees — leaves as well as large limbs — will also float on top of a puddle. Weight and water depth do not enter into this child's thinking, but the child is using his or her experience with trees and floating to make a reasonable prediction. Later experiences (observing part of a tree limb underwater) will allow the child to refine his or her prediction and the thinking behind it.

The ability to predict thus depends on how much experience young children have manipulating materials, studying their properties, transforming them in various ways, and observing the results. Moreover, simply encouraging preschoolers to make predictions gets them thinking in ways they might not do otherwise. Instead of focusing on the here-and-now (what is occurring in the present), they are being asked to think about the there-and-then (using what happened in another place and time to imagine what has not happened yet). The more opportunities children have to engage in this type of thinking, the better they get at reflecting on their experiences and observations and using them to make predictions (Church, 2003).

Teaching Strategies That Support Predicting

As children carry out their play themes and solve problems, they make predictions about how materials will behave, what their peers will do, and whether the solutions they try will work. However, this process may not be very deliberative or promote critical thinking unless adults help children actively reflect on what happened in the past to anticipate the future. You can use the following strategies to help children develop their capacity to make predictions.

Help children reflect on the similarities between their past and present experiences

Draw on the other components of the scientific method — observing, classifying, experimenting — to support children's emerging predictive abilities. Help them recall what they observed about the way familiar materials behaved to ask whether the same thing might happen again with similar objects ("Think about how you got the wood to stick to the paper yesterday at small-group time. I wonder if that might also work to attach the wood to the cloth").

At work time in the art area, Desmond asks his teacher Shannon to help him stick a craft stick to a piece of fabric. When Shannon asks Desmond to think about how he has made other things stick together, Desmond replies, "Oh, glue."

At work time in the house area, Jessie and Nina build a tent with four chairs, but when they drape a blanket over the top it falls into the middle. Jessie hands Nina a bed sheet and says, "Try this. It's not so heavy." When Nina successfully spreads the sheet over the chairs, Jessie says, "I knew it would work!"

Use preschoolers' interest in classification to help them determine whether another object or action fits in the same category as a familiar one and might behave the same way (e.g., they are both sticky and can be used to attach things). Alternatively, help them consider whether the new material does not belong in the same group and might therefore behave differently (it's heavy and may not balance on top). Perhaps a new object or situation is the same in some way (it's made of the same substance) but different in another way (it's bigger). Children may experiment to decide which data are relevant to making a prediction. Your role is not to tell children which aspect of their experience is the "correct" one to apply, but rather to help them be aware of similarities and differences and what they think are the implications of their knowledge.

Encourage children to say what they think will happen

Being concrete, preschoolers focus on the present, whereas prediction involves reflecting on the past and applying the lessons learned to form a hypothesis about the future. Asking children to make predictions helps them develop forward thinking as a "habit of mind." Since children do not always verbalize their predictions, sometimes adults must infer them from a child's behavior. For example, Nicole's plan is to paint a bunny and she mixes white and red paint. We might assume she predicts that mixing these colors will make pink. However, our inference may be wrong; perhaps those were simply the colors closest to her. Therefore, just as we encourage children to test their hypotheses, we should also verify our own assumptions. In this case, we can be more confident about Nicole's prediction if she states her intention to paint the bunny pink and her expectation that mixing red and white will make the color she wants.

KDI 48. Predicting

Teachers can support children's predictive abilities by drawing on other components of the scientific method. For example, adults can help children recall what they observed about familiar materials in the past, and ask if they think the same thing might happen with similar materials.

At snacktime, Jibreel goes to the window and says, "If it's not rainy today, then after large-group time and small-group time, we will go outside."

❖

At work time in the block area, Vadim says, "I think we'll need all the blocks to make the stage big enough."

❖

At outside time, Georgia says to Tyler, "Don't put in too much water or it'll get too muddy."

❖

At work time in the block area, Gabe and Dom get set to race their cars. "Mine's gonna win," predicts Gabe, 'cause it's bigger." His teacher repeats, "Your car will win because it's bigger?" inviting Gabe to say more. Gabe explains, "My uncle's truck is super big and it's way faster than my mom's car."

Encouraging children to state their predictions serves several purposes. As previously noted, it helps adults confirm whether children are operating on the basis of reasoned expectations (rather than trial and error) and scaffold

It's important for adults to encourage children to test out the accuracy of their predictions and ask them why they think their predictions did or did not come true. In this way they can reflect on the reasons behind their expectations and consider the actual outcomes.

their learning accordingly. More important, it creates an opportunity for children to think about and describe the experiences behind their prediction and why they think those experiences are relevant in the current situation. For example, when Rajiv predicts "There's gonna be lots of leaves at outside time" and his teacher asks why he thinks so, Rajiv points to the trees outside the window and replies, "It's really windy and the wind makes them fall down. I like to help my auntie rake leaves. She lets me jump in them." Furthermore, when children are able to voice their predictions, it opens the door for you to ask about the reasoning behind them and the conclusions they draw from their experiences (KDI 49. Drawing conclusions).

Encourage children to verify their predictions

Stating a hypothesis is one part of the scientific process. Checking out its accuracy is another key component. It seems logical to adults that when we make a prediction, we want to know if it is correct. However, this is not always the case with young children. For them, simply stating an expectation may be an end in itself. Having voiced their belief, children accept its truth. Adult encouragement, however, can also make them curious about whether their expectation matches reality. Then they will be ready to test out the accuracy of their hypotheses.

The same strategies that you use to help children verify their mathematical predictions also apply to helping them test their scientific predictions. Make comments and ask questions such as "Let's check it out" and "How can we be sure?" Discuss which aspects of the children's observations did (or did not) match their predictions. Ask them why they think their expectations did or did not come true. You don't have to know the answer yourself (so don't worry if you're not a "scientist"). What's important is encouraging children to check out their predictions and, based on what they observe, reflect on the reasons behind their expectations and the actual outcomes.

At work time in the block area, Shoshona says to Marie, "If you put that big one on top, it will fall down." Marie puts the big block on top anyway and the tower topples. Marie rebuilds it and this time accepts the small block that Shoshona hands her.

❖

At large-group time, Sanjay predicts that the wide, flat drum will make a louder noise than the narrow, deep one. His teacher encourages him to beat both drums and they listen together. Sanjay observes that contrary to his prediction, the narrow, deep drum is louder. When his teacher asks why he thinks that is so, Sanjay says, "It's taller so it has a bigger voice."

To see how children at different developmental levels make predictions, and how adults can support and gently extend their capacity to state and verify their hypotheses, see "Ideas for Scaffolding KDI 48. Predicting" on page 60. The chart offers ideas, in addition to the strategies described in this chapter, that you can use during play and other daily interactions with children.

Young children draw on their store of experiences to make an "educated guess" about an outcome.

Ideas for Scaffolding KDI 48. Predicting

Always support children at their current level and occasionally offer a gentle extension.

Earlier	Middle	Later
Children may	*Children may*	*Children may*
• When asked to make a prediction, indicate a choice as a way to answer a question rather than to make a prediction. • Show no interest in checking out predictions.	• On their own, make predictions at random (e.g., "This one, no that one, will go faster"); when asked to predict, make wild guesses. • Try out their predictions to observe if they are right or wrong (e.g., after predicting the round sled will get to the bottom of the hill before the rectangular one, try out both sleds).	• Make predictions based on prior experiences (predictions may or may not be correct); explain their predictions (reasoning may or not be correct) (e.g., "This snow pile will melt faster because it's in the sun"). • Try out their predictions and then confirm or change their thinking based on what they observe (e.g., after discovering the rectangular sled gets to the bottom of the hill first, say, "This time I think the rectangular sled will win again").
To support children's current level, adults can	*To support children's current level, adults can*	*To support children's current level, adults can*
• Accept children's choices (e.g., "You chose the red car"). • Clarify and repeat children's choices or predictions (e.g., "Tell me again which one you think will go faster"; "Marva says all the toys will sink").	• Repeat and acknowledge children's predictions (e.g., "You think that maybe it will rain today or maybe it won't"); accept that children may change their predictions. • Help children check out their predictions to see if they are right (e.g., "Let's see if mixing blue and yellow makes green like you think it will").	• Help children connect their predictions to their prior experiences (e.g., "You think the snow pile in the sun will melt faster because that's what happened yesterday"). • Repeat children's predictions and talk about what they observe; do not point out discrepancies but wait to see if they recognize them (e.g., "You thought the spiders would have two legs. What do you see under the magnifying glass?").
To offer a gentle extension, adults can	*To offer a gentle extension, adults can*	*To offer a gentle extension, adults can*
• Explain what children's choices mean (e.g., "You think the red car will go faster"). • Model checking out predictions (e.g., "I'm going to check to see which car is faster").	• Ask children why they think their predictions will happen (e.g., "What makes you think it's going to rain?"). • Ask children to describe what happened and if the result matches what they expected; label their idea a "prediction" (e.g., "You predicted blue and yellow would make green. What happened?").	• Encourage children to share and compare their predictions (e.g., "Todd thinks the sun will melt the snow faster. What do you think?"). • Ask children why they think their predictions did or did not happen (e.g., "The snowballs in the sun did melt faster. Why do you suppose that happened?").

CHAPTER 7

KDI 49. Drawing Conclusions

G. Science and Technology
49. Drawing conclusions: Children draw conclusions based on their experiences and observations.

Description: Children attempt to fit their observations and reasoning into their existing knowledge and understanding. They construct knowledge in their own way as they collect data to help them form theories about how the world works (e.g., "It's night because the sun goes to bed").

At small-group time, Josh puts his basket in a tub of water and then places toys in the basket. When it sinks, he says, "The rock makes it sink because it's so heavy."

At work time in the block area, Shelby tells Mark, "It fell down because you put the big block on top. It makes it tippy."

At work time at the sand and water table, after another child wonders if the snow outside will melt, Ella says, "The snow in the table will melt but not the snow outside, because it's still snowing."

At work time in the art area, Nicole tells her teacher, "If I wear a smock, I won't get paint all over my shirt. But I might still drip some on my shoe."

At small-group time, Davida sorts dinosaurs into two piles. She picks one up and says, "This is a T. rex. It doesn't fly because it doesn't have wings." She points to the other pile and says, "Those all have wings so they fly."

Drawing conclusions is the process of explaining what one has observed and incorporating it into one's current system of knowledge. Children make generalizations and draw inferences based on their previous experiences, making connections between what they are observing and what they already know. If their observations and the results of their investigations are consistent with their prior knowledge, they expand and generalize their existing concepts further. Jean Piaget (1950) called the process of taking in new information that fits with present knowledge *assimilation*. If there is a conflict between children's earlier beliefs and their current observations, they alter their thinking (though not always correctly) to take account of this new knowledge. Piaget called this process of adjusting one's thinking based on new information *accommodation*.

A conclusion can also be described as an interpretation of events. When young children draw conclusions, they make up a story about why something happened the way it did. But, whereas a literary story may be truth *or* fantasy, a scientific explanation strives *only* for truth. As young children explore a world of objects and actions, people, and events, they attempt to discover what that truth is. Just as scientists either prove or adjust their ideas based on the results of their investigations, children also change their thinking accordingly. The conclusions they draw in turn affect their subsequent behavior.

At work time in the book area, Brianna, Jayla, Ella, Jibreel, and Emily (the teacher) pretend to drive a car to California to visit their grandmothers. Brianna says to Emily, "They are not old grandmas because they don't have bumps on them."

At work time in the house area, Shironda puts on a floppy hat and dances over to the stove. When the hat falls off, she gets one with a narrower brim, but with her exuberant dancing, that one falls off too. She then puts on a small hat with a chin strap. This one stays on her head. At recall time she reports, "If you want to dance, you need a hat with a strap."

How Drawing Conclusions Develops

When preschoolers draw conclusions, they use all the scientific processes discussed earlier. They recall their observations, decide whether what they see is the same as or different from what they witnessed before (classifying), reflect on the results of their experiments, and consider whether the thinking behind their predictions matches the outcome or needs to be reconfigured. In other words, "When these pieces are all put together, these young investigators have the information they need in order to reach a conclusion" (Neill, 2008, p. 44).

In addition to applying the other components of the scientific method to reach conclusions and draw inferences, children also rely on other cognitive and social processes to create narratives about how the world works. Particularly important are the developmental capacities described in the KDI content area Approaches to Learning. For example, the wider the net of resources children cast in gathering information (KDI 5. Use of resources), the larger the database they have at their disposal to weigh the possibilities. Just as scientists require a big enough sample before they are confident of their conclusions (to make sure that what they observe is not merely a chance or fluke occurrence), so too do children profit from a sizable sample of evidence. If blue and yellow always combine to make green, it's a safe bet that the proposition "blue and yellow make green" is true.

Merely having experiences is not sufficient for reaching a conclusion, however. Being able to reflect on one's observations (KDI 6. Reflection) is also critical in interpreting how the world works. Many other individual and developmental characteristics, such as an openness to new experiences and a willingness to change one's thinking, also affect the ability and flexibility to draw conclusions. Further, voicing a belief about why things act as they do entails taking a risk. Therefore, all the factors that allow children to feel safe and secure in the learning environment can also affect their willingness and ability to arrive at conclusions based on their scientific investigations.

Preschool children draw conclusions based on observations that they generalize to similar categories or situations (Gelman, 1999). This ability to make inferences allows them to expand their knowledge base beyond immediate experience (for example, if roses smell and roses are flowers, then other flowers smell too).

At outside time, Sara and her teacher crouch to watch an ant. "Ants carry crumbs," Sara comments. "Do they carry anything else?" her teacher asks. "No, just crumbs," Sara answers.

Of course, breadth of experience alone may not be enough to change a child's conclusion. In fact, an erroneous theory can influence what they observe and lead to the further conviction of their error. They may actually misperceive something to fit their inference (if they think gooey things smell bad, they will sniff honey and declare it smells "yucky" rather than "sweet") or they may simply insist that their conclusion is right in the face of conflicting evidence. Focusing on their observations, rather than the "rightness" or "wrongness" of their conclusions, allows young children to change their ideas in a safe and risk-free environment. Also, after children have accumulated enough contradictory information, they are more inclined to adjust their thinking.

Just as scientists base their conclusions on a sizable sample of evidence, young children make inferences based on repeated experiences with a range of materials. If blue and yellow always combine to make green, for example, children can safely conclude that the proposition "blue and yellow make green" is true.

Young children's scientific conclusions are largely based on their perceptions and concrete experiences (Bruner, Olver, & Greenfield, 1996; Piaget, 1951). They pay attention to what is most obvious to them. That is why, for example, they do not yet conserve quantity. However, research shows that children can also make inferences about things they do not directly observe, including how bodies and minds work inside (Gelman, 1999). For example, preschoolers understand that "germs" exist, cause illness, and may be found in foods that appear clean. They develop "theories of mind" to explain how people's thoughts and feelings affect their behavior.

On the other hand, young children often overgeneralize within and between categories (for example, assuming that if one bird lays blue eggs, all birds lay blue eggs). They do not realize that a trait that is true in many cases provides a firmer basis for drawing a conclusion than a trait observed in a single case.

At greeting time, Jamal announces, "It snowed today because the weatherman said so. He should make it be warm."

At snacktime, Dillon says, "Vitamins make you grow strong. If you don't eat vitamins, your bones will break."

As the class heads to the play yard for outside time, Janey tells her teacher to turn out the lights. "It will save energy," she says, "and the world won't burn up."

Drawing conclusions is a complex process involving perception, cognition, and emotion. It depends on the opportunities in the learning environment, and the support and encouragement of adults. Despite — or perhaps because of — these complexities and challenges, young children are highly motivated to make inferences about how the world works and to base their actions and ongoing explorations on the conclusions they draw.

Teaching Strategies That Support Drawing Conclusions

To promote children's capacity to analyze information and draw conclusions, adults need to support their openness to discovery and ability to engage in critical thinking. The skills involved in making scientific inferences parallel the last two key developmental indicators in Approaches to Learning, namely making use of multiple resources (KDI 5. Use of resources) and reflecting on the data one collects (KDI 6. Reflection). The teaching strategies described below will help to foster these orientations in young scientists.

Provide materials and experiences that work in similar but not identical ways

To help preschoolers develop the capacity to draw conclusions, provide opportunities for them to work with materials and experience events that sometimes, but not always, act in the same way. Examples include play dough and clay, watercolor and tempera paint, and bristle blocks and magnet blocks. Look for opportunities outdoors as well. For example, children might observe that there are many types of bushes but only some grow berries. These experiences will allow them to make generalizations but also discover differences as they draw conclusions about them.

At outside time, Brent gradually adds more water to the sand. "The water makes the sand stick together," he tells his teacher, "so I can build walls."

At work time in the art area, Aaron tries attaching various things to his collage using the glue stick: glitter, sequins, beads, confetti, acorns, and metal screws. After the beads, acorns, and screws fall off the paper, Aaron concludes that "Glue sticks only stick little things."

At work time at the water table, Samantha says, "There's no more bubbles. We have to add soap." Connor replies, "No. We just have to blow more bubbles." He blows through his straw and makes a few bubbles, but not as many as there were at the beginning. Samantha squeezes in more soap and they blow until the bubbles reach the top of the water table. "We gotta do both," Connor concludes and Samantha nods in agreement.

To help preschoolers develop the capacity to draw conclusions from several sources of evidence, share your enthusiasm for looking at many examples and trying things in different ways. Make comments and ask open-ended questions to encourage them to gather a wide range of data and explore different alternatives. For example, as you work alongside children, you might ask one of the following questions:

- What else could you do (or look at) to find out?
- Is there another way to check out your idea?
- What about that other one? Do you think it will do the same thing or something different?
- Is there something else that looks (sounds, smells, etc.) like it? How about something that looks (sounds, smells, etc.) different?
- Let's try this one…and this one…and this one too. What do you see (hear, feel, etc.)?
- Are they the same? Different? The same in some way but different in another?

Encourage children to reflect on the processes and outcomes they observe

Drawing conclusions is a thoughtful process. Children are not only conducting observations and experiments, they are also thinking critically about what they mean. Encouraging preschoolers to reflect on what they observe, including the properties of materials and what happens when people and nature act on them, helps develop their capacity for higher order thinking and reasoning.

To promote thoughtful reflection, make comments ("I notice that the water isn't going down the drain") and ask open-ended questions that spur children to think about the things they observe or encounter ("Why does a heavy block on top make the tower fall down?"). You don't have to know the answers in order to encourage children to consider these questions. Remember that your primary role is not to impart a large body of scientific facts (which you might not know). Rather you are aiming to promote a spirit of inquiry in young minds and guiding children to use the scientific method in answering their own questions.

At recall time, when his teacher comments on the clogged glue bottles in the art area at work time, Oliver observes, "The glue didn't come out because there was dry glue in the top."

At outside time, Frieda says, "When I put the magnifying glass right on the rock it looks the same. When I stand back, it looks bigger." Her teacher asks Frieda why she thinks that's so. "Because there's more room for the rock to get bigger when I stand far away," she answers.

As children reflect on their experiences, it is particularly important to call their attention to disparities in and contradictions to their expectations. They may not notice these things on their own, but they may be intrigued by them if you point them out (e.g., "You thought the bigger one would weigh more, but it weighed less than the little one. Hmmm…").

At work time in the block area, Monte tells Jonathan, "We need four more [sections of train tracks] to get to the wall." Jonathan gets four sections but two are short pieces and the track still does not reach as far as Monte figured it would. "They all have to be the same size," Monte concludes. Jonathan gets two longer sections and they complete the track.

Finally, act delighted, surprised, and puzzled by what the world offers and the effects you and the children create in your investigations. A sense of wonder is contagious. So is curiosity. If you ask *what*, *how*, and *why* questions, young children are more likely to pose them too. In the process of figuring out how to answer them, and thoughtfully using the information generated, children will enhance their capacity for making inferences and drawing observationally based conclusions.

A sense of wonder and curiosity is contagious. When adults consistently pose what, how, *and* why *questions about materials and experiences in the classroom, children are more likely to ask these questions too.*

For examples of how children make scientific inferences at different stages of development and how you can scaffold their critical thinking in this area, see "Ideas for Scaffolding KDI 49. Drawing Conclusions" on page 68. The chart provides ideas for implementing the strategies described in this chapter as you play and interact with the children in your program at different times of the day.

Ideas for Scaffolding KDI 49. Drawing Conclusions

Always support children at their current level and occasionally offer a gentle extension.

Earlier	Middle	Later
Children may	*Children may*	*Children may*
• Learn about objects and events in the moment (e.g., balls bounce, roll, and can be thrown).	• Draw conclusions based on single observations or experiences; overgeneralize (e.g., all bugs sting).	• Draw conclusions based on multiple observations or experiences (e.g., some bugs sting and some bugs don't sting).
To support children's current level, adults can	*To support children's current level, adults can*	*To support children's current level, adults can*
• Call children's attention to what happens with materials and during experiences (e.g., what different things they can do with sand; what happens when they throw a feather down from the climber).	• Repeat or restate children's conclusions (e.g., "So you think all daddies are taller than mommies because your daddy is taller than your mommy").	• Encourage children to find more examples and ask whether they fit with or confirm their conclusion (e.g., whether pushing the space bar on *every* computer program restarts the game).
To offer a gentle extension, adults can	*To offer a gentle extension, adults can*	*To offer a gentle extension, adults can*
• Recall, and encourage children to recall, what they observed (e.g., "The dry sand ran fast through the funnel"; "What did the feather do when you threw it?").	• Challenge children's conclusions; encourage them to gather more information (e.g., "You think flowers smell pretty. Let's smell these other flowers too").	• Acknowledge when children change their conclusions to account for new information; ask them why they changed their conclusion (e.g., "So now you think only some flowers smell. I wonder why you changed your mind").

CHAPTER 8

KDI 50. Communicating Ideas

G. Science and Technology
50. Communicating ideas: Children communicate their ideas about the characteristics of things and how they work.

Description: Children share their questions, observations, investigations, predictions, and conclusions. They talk about, demonstrate, and represent what they experience and think. They express their interest in and wonder about the world.

At work time in the block area, Henry takes off his firefighter helmet, wipes his brow, and says, "Firemen wear sweaty hats."

At recall time, Madeline draws wavy lines on her paper and says, "That's the water in the water table."

At greeting time, Alana says as she takes off her mittens and puts them in her coat pockets, "Kangaroos don't need pockets cause they have pockets in their mommy's tummy."

At work time in the art area, Peter tells Jake, "I'm going to draw my truck with fire and smoke coming out because it's going really fast."

At outside time, Lakisha draws a flower in the dirt. She tells her teacher, "I made five petals just like this flower [holding up a flower she's picked]."

Preschoolers, excited by their scientific discoveries, are eager to communicate them with others. "As children explore, conduct investigations, and generate theories about how and why things happen, they are excited to share their results and do so in many ways — through spoken language, drawings, written words and symbols, demonstrations, gestures, and so on" (Neill, 2008, p. 17). Sometimes young children simply point to something as a way to communicate the message "This is interesting! Pay attention!" They may also talk about what they are observing with very little prompting. Merely encountering the world's wonders leads to spontaneous comments such as "Look what I saw!" "Did you hear that?" "You gotta feel this!" Getting others to focus on the phenomena that attract them is inherently rewarding for young children. The interest of adults and peers affirms for children that what interests them is worth paying attention to.

These simple gestures and remarks from children alert us to the fact that scientific awareness is taking place. However, it is during supportive conversations with adults, and in peer interactions facilitated by adults, that children are more likely to think about their experiences and develop theories about what they mean. Adults also help preschoolers reflect more deeply about their actions and observations by encouraging them to represent these events in other symbolic ways. They may use artwork, engage in pretend play, write or ask a teacher to write down their thoughts, and record data on simple graphs and charts. Children are more likely to elaborate on their observations, state and verify predictions, interpret what they discover through their senses, and generalize their theories when adults provide many opportunities for ideas to be communicated.

How Communicating Ideas Develops

Language is not only an indication of children's thoughts, it can actually shape their thinking and reasoning. Verbal communication serves several important functions in early scientific development. The more young children "talk" science, the more they will think "scientifically." For example, if children know a listener is eager to hear details about what they did and observed, they pay attention to multiple features of materials, actions, and outcomes so they have this information to share. If preschoolers hear and use "if/then" language, they observe and think about the sequence of events and cause-and-effect relationships. In this way, the very act of talking helps young children be more observant.

At work time at the water table, Anthony announces after he tries each material, "The rock sinks. The leaf floats. The paper floats. The bead sinks. The cork floats." He walks around the classroom looking for new objects to bring back to the water table and continues reciting his discoveries after trying each one.

At work time in the art area, Noelle paints a picture and asks her teacher to help hang it up. She points to the sun in her painting, saying "I mixed red and yellow and made orange."

Young children, like adult scientists, also use methods other than speech to communicate their ideas. They enjoy making drawings and models that show how they interpret the world. Chalufour and Worth (2003) say documentation is important in the formation of scientific concepts because it helps children find patterns and recognize relationships.

Children's representations also offer adults insights into their understanding of objects and how they work. For example, how a child sees a plant is reflected in the details of a drawing — does it include petals and leaves as well as a stem? If children accompany their representations with a narrative, it gives us further insight into their scientific understanding.

At small-group time, Joshua puts Lego blocks together and points out all the parts of his "jet plane" to his teacher. "There are motors on the back and this is where the pilot sits and these are the wings, and it has jet shooters here and this is the cab."

At work time in the block area, Anthony explains to his teacher the structure he built with Lego blocks. "That's an explorer car. They go to where the wild animals are. The dinosaurs are the most wild animals ever. The explorer car has to be very fast to get away from them."

Preschoolers' role play also reflects their understanding of how the world works. For example, if they twist the oven dial to bake circles of play dough arranged on a tray, they are communicating nonverbally their knowledge that the stove has to be on to heat up the batter to make cookies. The stories they dictate to adults to write down often include their scientific observations.

At work time in the block area, Jason says, "Everybody get away from the dynamite. I feel it shaking. The earthquake's blowing up higher than a tree."

❖

At work time in the block area, Mikki makes a car with blocks and uses two small blocks as the accelerator and brake. She says, "You have to press down real hard to make the car go fast."

❖

At recall time, Pascal draws a picture of the fort "garage" he made for his Matchbox cars using small blocks. He draws thick lines where the roof meets the walls, and asks his teacher how to spell "tape." He explains that he used tape so the roof blocks would stay up.

As their skills in data analysis (KDI 39) develop, preschoolers can also — with adult help and support — use simple charts to represent their scientific ideas.

At small-group time, the children use magnets to attract different objects. With their teacher's help, they make a chart labeled "Sticks" and "Does Not

Children enjoy making drawings and models to communicate their understanding of scientific concepts and to show how they interpret the world.

Stick." They discuss what is the same and different about the objects in each column.

Thus, just as listening to their own speech helps children reflect more actively on what they did and learned, so can creating and examining various forms of representation spur them to consider their observations in greater depth. For example, children may think about the details they observed (color, size, texture) in order to choose appropriate materials to represent these attributes in a drawing or model. They decide which props to use or what verbal directions to give a playmate to reflect their understanding of how something works during pretend play (for example, what types of treatment the doctor has to give the sick baby to make it well again). In these ways, representation helps to deepen children's experience and communicate their thinking to adults.

Teaching Strategies That Support Communicating Ideas

The use of language and opportunities to see ideas represented in books, pictures, and other ways, helps to deepen children's understanding of science (Worth & Grollman, 2003). To support preschoolers as they communicate their observations and ideas about the natural and physical world, you can use the verbal and nonverbal teaching strategies described below.

Use scientific language as you talk with children about their actions, observations, and discoveries

Science is interesting and children need interesting language to communicate the full range and excitement of their discoveries. To help preschoolers talk about their observations and conclusions, use the same strategies that encourage them to speak in general (KDI 22. Speaking), and apply these ideas to the vocabulary of science.

First and foremost, encourage children to talk about their scientific experiences in conversations with you and their peers. Science learning, like most other learning at this age, takes place in a social context. As Christopher Landry and George Forman (1999) note, "Science, for children as well as adults, is not done in a vacuum but in a social realm within which ideas are discussed, debated, and take shape" (p. 137). Recall and snacktime provide many opportunities for these discussions. Listen to what the children say and also share your own observations and discoveries with them. For example, at greeting time you might introduce the word "frost" as you talk about having to scrape the ice off your windshield to get to school.

Provide a secure and risk-free environment within which these exchanges can occur. Children should feel safe sharing their speculations without fear of criticism or ridicule. Invite and accept children's ideas. As in the other areas of science education, your role is not to correct children's thinking or supply unrelated facts, which they are not developmentally ready to absorb anyway. Rather, your goal in communicating with children about science is to make comments and ask open-ended questions that help them evaluate what they see, make connections between their experiences, and construct explanations that make sense within their current system of logic.

Even more important than what you say is encouraging children to ask their own questions, raise doubts, and voice surprise when their observations do not match their expectations. Let children know you are interested in what they have to communicate with gentle questions that ponder "what, where, when, how, and why." Wait patiently so children have time to frame their answers and think about your comments and questions.

At work time in the toy area, while holding a pine cone, Ella explains, "They fall off the tree and then they are on the ground in the grass." Her teacher asks what happens after the pine cones are on the ground. Ella says, "They grow new trees," and after thinking for a moment adds, "But you have to carry them and plant them."

Building a vocabulary to talk about science takes time. Use a variety of words repeatedly, so children will eventually understand and use these terms themselves. In addition to the language you introduce, encourage children to describe what they sense in their own words and incorporate their words in your communications. For example, in the following anecdotes, the teacher repeats the child's language and then introduces a new word or idea.

At work time in the block area, when Tyne suggests using the "tunnel block," his teacher says, "Tyne thinks we should use the 'tunnel block.' That's the block that's shaped like an arch."

❖

At greeting time, Melody tells her teacher that it's "smoky" outside. Her teacher replies, "It does look smoky outside. That's fog."

❖

At outside time, Jason says, "The puddles are gone." His teacher says to the other children gathered on the pavement, "Jason noticed the puddles are gone. The water in the puddles evaporated."

Since science begins with observation, use descriptive words yourself as you talk with children about objects and events. In addition, introduce the language of scientific reasoning. Remember that science learning can occur in virtually any context. So when you talk about children's daily activities, use words such as *if* and *then* to help them observe and express ideas in terms of cause and effect ("You saw that if you mix blue and yellow water, then you get green water"). Include the word "because" in conversations to get children thinking about *why* something happens ("You think it fell down *because* the block you put on top was too heavy") and *how* materials and events are related ("It's snowing *because* it's cold outside").

Provide opportunities for children to symbolically represent their scientific experiences

In addition to sharing their ideas through spoken language, young children use other means of communication to represent their scientific understanding. They do this when they create artwork, engage in pretend play, and "write" or dictate stories and songs. With adult support, they may also represent their ideas on simple charts and graphs.

Artwork. Children spontaneously use art materials at work time to represent the things they observe and their understanding of how the natural and physical world works — how houses are built, what animals look like and do, and what makes things go fast. Asking children to describe or explain their artwork gives you a window into their scientific interests and thinking.

At work time in the art area, Lily makes circular scribbles on her paper and explains to her teacher, "This is a tornado. It's a bad storm that blows things away."

❖

At work time in the toy area, Piper asks her teacher Sheila to take a photo of the kitty house she made with Magna-Tiles. "I want to show it to my daddy," she explains, "because it's like the one he built for Izzy [the family cat]." Sheila takes the photo, then kneels down to look more closely at the structure. "Tell me how it's like Izzy's kitty house at home," she asks. Piper points and explains each part: "This is the door. You put his food here [she points to the smaller window] and his water through this one [she points to the larger window]."

At work time in the toy area, Josie puts shells and marbles on the balance scale. She gets paper and markers and tells her teacher, "I'm gonna draw a picture with two shells here [she points to one side], a line down the middle, and six marbles here [she points to the other side]." Her teacher comments, "You are going to draw a picture of how you got the shells and marbles to balance on the scale."

You can also use art to help children communicate at recall time, during small-group time, or as follow-up to a field trip. When preschoolers draw or make models to represent the materials they used, the actions they performed on them, and how things changed as a result, it is a way for them to share with others, "This is what I saw, here is what I did, and this is what happened." Creating art not only helps children communicate their scientific ideas, it also provides an opportunity for them to reflect on the materials and processes they are representing.

Pretend play. Children act out their understanding of how things work when they role-play individually and with others. For example, when children in the house area mold play dough for a birthday cake, they may consider how thick it needs to be to hold up

In communicating with children about science, the teacher's goal is to offer comments and ask open-ended questions that help children make connections between objects and events and construct explanations that make sense to them.

Popsicle sticks as candles. "Blowing out" the candles communicates the idea that "wind can put out fire." Later on, they build a "campfire" out of straws and pipe cleaners, then blow on it because "wind helps get fire going." In another scenario involving dress-up clothes, a child may give a playmate a heavy jacket to wear when they play "snow explorers" because they are going some place "very cold."

To support and extend children's ideas about the nature of materials and scientific processes, provide time, props and prop-making tools, and participate in the play scenarios they create. For example, while pretending to make a cake in the house area, a teacher asks for something

Creating art not only helps children to communicate their scientific ideas, it also gives them the opportunity to reflect on the materials and processes they are representing.

to stir with. A child hands her an electric mixer (with the cord removed) and says, "You can't really plug it in. You have to pretend it works." To extend the child's idea, the teacher replies, "Yeah, you need a cord so electricity can run the motor that turns the beaters."

Writing. By providing writing tools such as clipboards and markers throughout the classroom, as well as during group-time activities and field trips, you can encourage children to record their ideas. Reading books about science with children will also inspire them to communicate their ideas by "writing" communications of their own.

At work time in the house area, Rene gets an index card and pencil and writes "1" (for one cup of flour) and "3" (for three cups of sugar). She asks her teacher how to write "5" (for five spoons of chocolate chips) and copies the numbers onto the card. She explains, "This the recipe for chocolate chip pie so I'll remember to make it the same way tomorrow." She puts the card in the recipe box.

Lists, charts, and graphs. Data analysis in mathematics and science are closely linked. Young children can apply their emerging ability to create and interpret simple quantitative records to share the results of their scientific investigations. As you support preschoolers in the process of documentation (i.e., pictures, words, simple graphs and charts), you are helping them review the steps in the preschool scientific method. For example, making a simple chart helps them reflect on what categories to create ("How should we label the columns?"), what symbols to use to annotate the information ("Should we make hatch marks or write numerals?"), and most significantly, how to interpret the results ("I wonder why there are more in this column than in the other column").

During work time in the art area, Colleen makes a chart of the animals that live in her house. She writes numerals in one column and draws pictures of the animals in the other column.

Preschoolers vary widely in their ability to communicate their ideas about science. For examples of how children at different developmental levels share their observations and conclusions, and how you can appropriately scaffold their learning, see "Ideas for Scaffolding KDI 50. Communicating Ideas" on page 78. Use the suggestions in the chart, and the strategies described in this chapter, as you play and interact with the children in your program.

Sometimes young children point to something they've created or observed as a way to communicate to others, "This is interesting! Pay attention!" These simple gestures and remarks alert us to the fact that scientific awareness is taking place.

Ideas for Scaffolding KDI 50. Communicating Ideas

Always support children at their current level and occasionally offer a gentle extension.

Earlier	Middle	Later
Children may	*Children may*	*Children may*
• Call attention to their simple scientific observations and discoveries (e.g., point excitedly when a bird takes flight; say, "Hey look!" when they splash their foot in the mud). • Demonstrate their knowledge of the physical and natural world through imitation (e.g., flap their arms like a bird).	• Provide a simple verbal description of what they see or do (e.g., "Oh! The bird flew out of the bush!"; "Ooh. My foot sank in the mud."). • Make simple representations (such as drawings or models) of their scientific discoveries (e.g., paint a "flower" with a stem; mold hills and rivers out of sand and water).	• Provide a verbal explanation of how or why something happens (e.g., "I think the bird flew out because she needs to get food for her babies"; "This mud is gooey. It made my foot stick to it"). • Make complex representations (such as detailed drawings or models, dictation or their own writing, marks on a chart) of their scientific discoveries (e.g., draw a block tower to show the smallest block goes on top; make signs to indicate "hot" and "cold" for each water table; make a mark in a column to indicate that the cotton ball is fuzzy).
To support children's current level, adults can	*To support children's current level, adults can*	*To support children's current level, adults can*
• Share children's excitement in their scientific discoveries (e.g., stand side by side to watch the bird fly off together); attach words to their gestures (e.g., "You're making the wet mud splash"). • Imitate children's actions; comment on what they are doing (e.g., "I'm flapping my arms like a bird too").	• Acknowledge and repeat children's observations (e.g., "Yes, the stem has prickers"; "That bird did fly out of the bush!"). • Provide materials that children can use to represent their scientific discoveries (e.g., art materials, construction toys).	• Listen to children's explanations; give them time to think about and express their ideas. • Ask children to describe or explain their complex representations (e.g., "Tell me about your picture"; "What does the check mark on this chart stand for?").
To offer a gentle extension, adults can	*To offer a gentle extension, adults can*	*To offer a gentle extension, adults can*
• Encourage children to tell you what they observe (e.g., "What did you hear when the bird flew out of the bush?" "What happened when you jumped in the puddle?"). • Add another idea to children's imitations (e.g., "I'm going to pretend to peck worms out of the ground").	• Wonder why something that children observe happens (e.g., "I wonder what made the bird fly out of the bush so suddenly"). • Help children focus on additional details they can include in their scientific representations (e.g., provide actual objects, photos, and illustrations of things in the natural and physical world).	• Ask questions to encourage children's scientific reasoning (e.g., "I wonder *what* makes tape sticky"; "*How* does the energy get into your bones?" "*Why* does cheese make your bones stronger than candy?"). • Encourage children to seek out additional information to include in their detailed representations (e.g., to think of another property of fabric and add a column for it on a chart).

CHAPTER 9

KDI 51. Natural and Physical World

G. Science and Technology

51. Natural and physical world: Children gather knowledge about the natural and physical world.

Description: Children become familiar with characteristics and processes in the natural and physical world (e.g., characteristics of plants and animals; processes of growth and death, freezing and melting). They explore change, transformation, and cause and effect; they become aware of cycles that are meaningful to them.

At outside time, Cory tells Sam, "Bees get nectar from flowers. They need it to make honey."

At work time in the house area, Kovid and his teacher Sue talk about sharks. Kovid says that "Hammerhead sharks have something like hammers on their heads." He adds that sharks live in oceans and can "smell blood from very far away, one pint in a lot of water."

At small-group time, Matthew puts Legos together and explains, "This is my rocket and these are the motors. They make it take off and stay up in space."

At outside time, Marlene tells her teacher Tammy, "Moths go to sleep in little houses like sleeping bags and when they wake up they are butterflies."

At planning time, Marva says she is going to bake bread in the square pan. "If you bake it on a round plate, it drips over the side. My mommy did that and it burned up the oven."

At outside time in the sandbox, Lucius shows Tony his monster truck. "It uses sharp things sticking out of the wheels so it can climb up," he explains.

Early science education emphasizes the scientific method rather than expecting children to learn specific facts. Nevertheless, there are certain areas of knowledge that preschoolers can appropriately begin to get acquainted with. This information builds on the children's own interests and their personal observations of nature (their own bodies, plants and animals, weather patterns) and how the physical world works (how machines operate, why things of the same size have different weights, why certain materials change color or produce bubbles when mixed together, why raising one end of a ramp makes cars go down faster). When they use the preschool scientific method to explore and interpret their experiences with these materials and processes, young children construct theories about the natural and physical world.

How Knowledge About the Natural and Physical World Develops

The environmental features and events that capture a young child's attention and curiosity are virtually unlimited. Listen to the questions that children ask to find out what interests them. For example, here are some of the questions overheard in one preschool classroom over the course of a year:

- Why is the stick heavier than the feather even though they are the same size?
- How does the sun light up the sky?
- What makes a shadow?
- Why do ice cubes melt?
- Why can I slide on ice but not on wood chips?
- What do plants eat?
- Why did my dog die?
- How do scissors cut?
- What makes gears go around?
- Why did the bulb burn out?
- Why does a magnet attract some things and not others?
- Where does the food go after I swallow it?
- Why does medicine make us better?
- Why is your hair a different color than mine?
- Why does sand stick to my skin?
- Why do rocks come in so many different sizes?
- Why does it snow when it's cold outside?

Because so much of the world is new to them, young children are constantly collecting data about the natural and physical environment. They take in vast amounts of new information every day. Eventually, they need a way to store all this input. When preschoolers ask "why" and "how" questions, they are demonstrating their interest in understanding and organizing this information.

Young children bring to science many ideas about how the natural and physical world works. Although their concepts seem "naive" to adults, they are often quite sophisticated when viewed from the perspective of the child's system of logic ("The sun makes things warm. If the sun's out, it must be warm, even in the winter"). Children hold on to their theories because they make sense to them and serve a useful organizing purpose. They may therefore not be receptive to alternative ideas offered by adults. According to Landry and Forman (1999), "The child who is told a new explanation will often concurrently hold the misconception because it is more integrated into many other assumptions he has developed through experience in the world" (pp. 138–139).

At outside time, Josh explains to Chandra how to make mud. "First it's sand. Then you add water and it's mud."

At work time in the block area, Trey shows Douglas what he is making. He rotates a small block on top and explains, "It's a machine that twists and gives you candy flavors." Douglas points to the twine he has wrapped around a large block and tells Trey, "I'm making an exercise machine that strongs up your legs. You pull the string tight and it makes muscles."

At work time in the book area, while looking outside at the snow, Joey says, "Snow is water. Water is in the snow. When you put your hand out, the snow goes on it and then the water comes out."

Children do not merely observe the world and store the information as "raw data."

Scientific knowledge is ultimately about drawing conclusions, and children engage in this process all the time. As explained earlier, they "assimilate" compatible information to confirm existing theories and "accommodate" conflicting information by altering those theories, although not always correctly (Piaget, 1950; 1955). Developmental research shows that preschoolers tend to reason by analogy. They recall a similar experience (comparable materials or events) and generalize to the new situation that confronts them (Russell, Harlen, & Watt, 1989). This may lead to the wrong conclusion, but its roots are traceable to the child's experiences and system of logic. For example, young children may wonder why a dish containing water earlier in the day is now dry. Based on their experience with sponges, they might conclude that the dish has absorbed the water. Or, having seen wet pavement dry in the sun after a rainstorm, preschoolers might reason that the water "floated" up into the air. In other words, their explanations about the natural and physical world make sense given their experience. And they are highly motivated to construct an explanation that resolves what they see with what they already know or believe.

The sand and water table offers children the opportunity to explore the features and processes of the natural and physical world.

Teaching Strategies That Support Learning About the Natural and Physical World

Young children are curious about the attributes and actions of things in the living world and the features and mechanical processes of objects and events in the physical world. To support these interests, adults can use the following teaching strategies when interacting with children in the indoor and outdoor learning environment.

Provide materials and experiences for children to gather knowledge about the natural and physical world

Children's thinking progresses from the concrete to the abstract. Therefore a good starting place to help preschoolers construct ideas about the world is to assemble collections of living and nonliving things they can compare (plants, animals, shells, seeds, salt and fresh water, soil, rocks, hardware fasteners). Supplement real objects with printed materials (books, photos, catalogs) as well as realistic replicas (small wooden or plastic animals, artificial flowers) of things that cannot be gathered (they are too large, not available, or would damage living things if collected). Encourage children to sort, handle, and describe these collections to discover similarities and differences in their properties. Encourage them to use all their senses to observe whether and how materials change over time or as a result of their actions (a wet towel is heavier than a dry towel; colored leaves turn brown and brown leaves crumble if you squeeze them; flowers decay and lose their smell). For more on making and describing collections, see classifying (KDI 46. Classifying).

At work time in the art area, Tasha talks with Sue (the teacher) and Senguele about the movie "Madagascar." She explains, "'Madagascar' has lions, zebras, and penguins. It's a movie." Sue asks if there are any other animals in the movie. "No," explains Tasha. "It's the jungle. Bears need to live in the forest, so they can't go to Madagascar."

At small-group time, Hannah explains to her teacher Johannes how the garlic press she is using with play dough works. "Close it and push it. Then it comes out like this." Johannes copies what Hannah does and then comments, "I wonder what else the garlic press will work with." Hannah surveys the other materials on the table and selects the modeling clay. "You have to push that harder," she concludes. Then she tries a piece of felt, and observes that "it makes bumps but it doesn't come out."

Use objects and experiences to build children's awareness of basic animal and plant needs (air, food, water, light, rest); natural and urban environments, especially those found in the local area (bodies of water, forests, beaches, hills, farmland, meadows, deserts, snow-covered mountains, towns and cities, parks, playgrounds); and cycles such as birth, growth, and decay and changing seasons and weather patterns. Remember that weather and seasons in the abstract hold little meaning for preschoolers. However, they are interested in current conditions they directly sense and short-term changes that affect their day-to-day activities (for example, whether they can play outside, what kind of clothes they wear, whether there will be rainwater to make mud or snow to build a fort, if they are likely to find worms, whether frost has killed the remaining plants in the garden).

Take field trips around the neighborhood and visit farms and orchards, ranches, fisheries, zoos, pet stores, safe building sites, hardware stores, planetariums, aquariums, the beach, natural history museums, botanical gardens, and Earth Day events. Talk with owners, tradespeople, and museum docents beforehand to make sure the experience is appropriate for young children, (i.e., they can touch things, ask questions, have space to move, and so on). While you are there, take pictures to put in a classroom album so the children can look at the photos and talk about their experience. Bring back materials the children can sort and use as props in their pretend play. Follow up the experience with relevant small- and large-group times, planning and recall strategies, and transitions. For example, after a trip to a pet store, add empty boxes of dog biscuits and dog dishes to the house area, books about pets to the book area, and small animal figures to the toy area. Following a trip to a farmers' market, plan a small-group time in which the children peel and husk ears of corn, and explore the different parts of the plant (kernel, cob).

Encourage children to make connections to explain how the world looks and functions

Use concrete examples and everyday experiences to help children consider the meaning behind the attributes of things. Build on the objects and events they gravitate to naturally in the indoor and outdoor environment. For example, ask "how" and "why" questions that lead children to reflect on the reasons for something's appearance or movement (e.g., "I wonder why flowers come in different colors"; "Why are some stones flat while others are round?" "I wonder how birds fly"; "Why does that door swing closed by itself?" "Why does pumping your legs make you go higher on the swing?") You do not have to know the answers to these questions yourself. In fact, you may puzzle over the same issues as the children. Part of the fun of science is generating theories and seeking answers together with them.

As children speculate and offer explanations, acknowledge and comment on their observations and theories. Refer them to one another to broaden their perspective and expand their thinking. As noted earlier (Campbell, 1999), children are more inclined to construct their own theories after hearing what their peers think. They are also more likely to examine and revise their reasoning when a peer rather than an adult calls their attention to contradictory evidence.

At work time in the toy area, Brianna holds a shell up to her teacher Emily's ear and says, "Listen to the ocean. Shells come from the ocean." Emily asks how the sound from the ocean gets into the shell and Brianna replies, "The waves carry it inside. When the water dries, the noise is left." She holds the shell to her own ear and nods as if to confirm her explanation.

At greeting time, Ian says his family is making an ice rink in their backyard. "We're going to shovel snow, put water on it, and wait one minute for it to turn to ice." His teacher comments to the group, "I wonder how snow turns into ice." The children offer various theories: "When the water dries, the snow gets hard"; "The water makes the snow cold"; "It takes more than a minute. You gotta wait all night"; and "When the water freezes, it gets slippery."

Adults can also use comments and occasional open-ended questions to encourage children to think about how things function and act upon one another. Take advantage of

Teachers can follow up class experiences with relevant activities, such as this small-group time following a field trip to a local apple orchard.

children's sense of awe and wonder to delve deeper into considering why and how things work they way they do. For example, if children are fascinated by the worms living under a rock they have turned over, ask them about the features of that environment and how it supports the worms' survival — "It's dark under the rock. I wonder what they will do now that we let the light in. Let's watch them and see what happens."

Preschool children ask adults many "how" and "why" questions. In supporting the development of young scientists, our goal is to help them use evidence and reasoning to answer their own questions. Your comments and open-ended questions will establish a spirit of inquiry in the classroom that moves children from asking others to creating their own explanations. Their thinking will progress from making increasingly detailed observations to constructing theories about how and why things in the natural and physical world work.

For examples of how children demonstrate their understanding of the properties and processes in their environment, and how adults can support and gradually extend their thinking at different stages of development, see "Ideas for Scaffolding the Natural and Physical World." Incorporate these ideas into your daily play and other interactions with the children.

To help children construct ideas about the world, adults can provide concrete examples of living and nonliving things children can explore and compare.

Ideas for Scaffolding KDI 51. Natural and Physical World

Always support children at their current level and occasionally offer a gentle extension.

Earlier	Middle	Later
Children may	*Children may*	*Children may*
• Name objects and events in the natural and physical world (e.g., rock, flower, rain). • Focus on the state of an object or event in the moment rather than notice a change (e.g., "The banana is mushy"; "I'm cold").	• Notice and/or comment on attributes in the natural and physical world (e.g., "Birds have feathers and fly"; It's hot when the sun comes out"). • Describe a change in an object or event (e.g., "It's not cold like yesterday").	• Compare and contrast attributes in the natural and physical world (e.g., higher ramps make cars go faster; plants differ in height). • Explain that a change happens because of something else (cause and effect) (e.g., "It's not as cold as yesterday because it's sunny out").
To support children's current level, adults can	*To support children's current level, adults can*	*To support children's current level, adults can*
• Repeat the names children assign to objects and events in the natural and physical world; if children use the wrong label, use the correct one. • Acknowledge children's observations about objects and events in the moment (e.g., "The banana *is* mushy"; "I'm cold too").	• Observe and comment alongside children (e.g., "The petals on this flower are red and the petals on that flower are yellow"; "The sun is really hot today!"). • Provide opportunities for children to observe materials (e.g., fresh flowers) and carry out activities (e.g., construction projects) that change over time; help them recall the change (e.g., document it with photos; ask them to describe the changes at recall time).	• Encourage children to observe more objects and processes to make more comparisons (e.g., compare the taste and juiciness of different fruits at snacktime; see how far they can throw balls of different sizes). • Restate children's cause-and-effect explanations; use *if-then, so,* and *because* language (e.g., "The sun is out *so* it's warmer today").
To offer a gentle extension, adults can	*To offer a gentle extension, adults can*	*To offer a gentle extension, adults can*
• Label additional objects and events in the natural and physical world (e.g., toad, mountain, stream). • Describe and label other states of the same objects and events (e.g., "Sometimes bananas are hard"; "Yesterday I was too hot!").	• Provide materials and experiences that help children make comparisons in the natural and physical world (e.g., provide odd-shaped and regular-shaped blocks; visit a farmers' market). • Encourage children to consider what caused the changes they observe (e.g., "I wonder why the flowers are drooping today").	• Ask children to explain the contrasts they observe (e.g., "I wonder why some fruits are juicier than others"; "Why do you think you threw the small ball farther than the big ball?"). • Pose "What if…?" questions, building on children's cause-and-effect ideas (e.g., "What if the sun never went down?").

CHAPTER 10

KDI 52. Tools and Technology

G. Science and Technology
52. Tools and technology: Children explore and use tools and technology.

Description: Children become familiar with tools and technology in their everyday environment (e.g., stapler, pliers, computer). They understand the functions of equipment and use it with safety and care. They use tools and technology to support their play.

At work time in the block area, Theo asks Natalie (another child) to help him make the flashlight work. She checks to make sure the switch is in the "on" position. Then she says, "Sometimes you have to just shake it," which she does and the light goes on!

At work time in the woodworking area, Lyle puts on safety goggles and uses a hand drill to start a hole for a nail in a block of soft wood. He puts the nail into the hole, then says he has to make the hole "more deep." After he drills deeper, he balances the nail in the hole and pounds it in further with a hammer.

At outside time, Cameron attempts to scoop sand into a tall, narrow container but most of the sand spills down the sides. He looks around and sees a funnel, which he places in the mouth of the container. "There," Cameron says, as he successfully pours in the sand.

At work time in the house area, Carol sits in front of the computer screen and Simone sits beside her as they make printed menus for their restaurant. They decide to serve macaroni and soup. Simone tells Carol to type M for macaroni and S for soup "like in my name." They print M and S pages and open their restaurant for business.

Young children use many kinds of tools as they make plans and carry out their intentions. For example, they may staple strips of fabric to paper to create a collage, cut pieces of yarn to make hair for a puppet, hammer golf tees into Styrofoam, tape a wad of newspaper to the end of a tube to make a magic wand, look at the leaves of a plant under a magnifying glass, or use a wagon to transport large blocks across the room. Computers and other technology are tools preschoolers can use to carry out their play ideas, acquire knowledge and skills, or solve problems. For example, they might use a drawing program to mix colors onscreen, see how many ways they can flip or turn a geometric shape, or use a simple word processing program to type and print a party invitation. Using tools and technology is thus an interesting end in itself (mastering the skills needed to use a stapler, or pushing the buttons on an audio player) as well as the means to an end (stapling the ends of two pieces of ribbon together to make a longer leash, listening to a favorite song).

How Use of Tools and Technology Develops

While developing the dexterity to use tools is an aspect of physical growth, understanding how tools work and using them to carry out investigations is part of early science learning. Problem-solving with tools builds young children's conceptual awareness because it involves "planning, sequential thinking, and predicting what specific actions with a tool might do" (Haugen, 2010, p. 46). Using tools also requires flexible thought and action, whether to use the tool itself with greater dexterity or to choose the tool(s) that will best accomplish one's goal.

Experimenting with tools

Children initially engage with tools to explore them as materials in their own right. For example, they may be fascinated by the feel of holding and pounding a hammer, the stickiness of tape or the motion of stretching it out from the roll, the motion of stirring with a spoon, or the satisfying "clicking" sound made by depressing a stapler. As preschoolers experiment and become more adept at using tools, they increasingly focus on the effects of their actions (tape sticks pieces of paper together as well as your fingers; stirring with a spoon mixes paint colors together).

Eventually, as children gain familiarity and competence with various tools, they contemplate their potential uses. They generalize from a tool's applicability in one setting to its relevance in another context. They consider how a tool can help them accomplish a goal or solve a problem.

Furthermore, preschoolers can now draw on their cumulative experiences with a tool to create a mental representation of it. This ability allows them to imagine how the tool can be applied in a new situation, the various ways it can be manipulated, and the result(s) it can produce.

At work time in the book area, after using a magnifying glass at outside time the previous day to examine a squashed spider, Davida holds the glass above a sheet of newspaper "to make the letters look bigger."

At recall time, Henry shows the other children a "hose" he made by attaching string to a block. He says, "I had to use this [electrical] tape to make it stick. That [cellophane] tape didn't work."

Once children have made the step from exploration to application, they increasingly use tools in the service of something else, such as making props to carry out pretend-play ideas, observing and experimenting, and solving problems with materials. Of course, like adults children encounter new tools or discover new uses for familiar ones, so exploration never ceases. Moreover, the ability to understand how tools work continues to mature along with their cognitive development.

Using computers and other technology

Technology can play a useful role in early science (as well as mathematics) learning when used appropriately. That is, it should supplement rather than replace hands-on learning with real materials (Hyson, 2003). Research shows that computers work best when (a) used in moderation and (b) coupled with suitable off-computer activities. For example, Susan Haugland (1992) found that children using developmental software alone gained in nonverbal skills, memory, and dexterity. However, those who worked with both the developmental software and freely chosen supplemental activities gained in these areas, but also in grew in

Children can use computers and other technology to carry out their play ideas, acquire knowledge and skills, or solve problems. Research shows that computers work best when used in moderation and when they're used with related off-computer activities.

verbal, problem-solving, and conceptual skills. They also spent less time at the computer, preferring the other self-chosen activities. By contrast, children using drill-and-practice software spent the most time at the computer but showed less than half the gains of the other two groups.

Young children benefit from becoming familiar with the mechanics of hardware (turning a machine on or off, using arrow keys to navigate the screen) and learning to use a few appropriate software programs. These programs should be interactive, open-ended, and promote discovery learning. Good software poses a problem, asks children to solve it, and provides instant feedback on their responses (Clements, 2002). Programs that pose problems with "correct" answers can be productive if the feedback causes children to reflect on their reasoning and attempt to solve the problem differently. Adults working alongside children can also provide this type of support.

In addition to promoting reflection and self-evaluation, technology has the added advantage of increasing children's manual flexibility and use of other hand-held tools. Using a touchpad or keyboard, children can sometimes move objects onscreen more easily than real objects that are hard to hold or maneuver (rolling a boulder big enough to cover the opening of a cave). While computers can never substitute for the sensory feedback provided by the actual materials, they can extend the range of objects and movements children can experiment with. Further, "contrary to initial fears, computers do not isolate young children. Rather they serve as potential catalysts for social interaction" (Clements, 1999, p. 122). The more open-ended the software, the more collaborative and less competitive children's play. Observations show that preschoolers enjoy working together at the computer where they solve problems, talk about what they are doing, help and teach friends, and create rules for turn-taking and cooperation. In fact, "young children may approach technology in this way; they may find it less intimidating when approached as a group project" (Elkind, 1999, "Observing Young Children Learning").

Teaching Strategies That Support Use of Tools and Technology

Children are curious about tools and technology in their own right, and also as resources to help them carry out their play ideas, gather information, and solve problems. The following teaching strategies will help you support their growing understanding of how tools and technology work.

Provide a variety of tools in all areas of the classroom

Because tools are resources for all types of scientific discovery, it is important to include them in every area of classroom. Moreover, don't be restricted to what we typically think of as "science" tools (e.g., magnifying glasses, binoculars, prisms, magnets, balance scales, wheels, pulleys, gears, kaleidoscopes, plastic thermometers, mirrors). The following tools commonly found in other parts of the indoor and outdoor learning environments also encourage children to explore how they work and to experiment with the ways they can be used.

Art area tools — Paint jar pumps, Popsicle sticks, plastic spoons, brushes, ink pad and stamps, stapler, hole punch, scissors, plastic molds, tape, glue (paste, liquid glue, glue sticks), string, yarn, ribbon, buttons, large-eyed tapestry needles, pipe cleaners, rolling pin, wire or canvas mesh, fabric scraps, straws, paper clips

House area tools — Kitchen utensils (wooden spoon, spatula, ladle, whisk, funnel, scoop, garlic press, measuring cups and spoons, timer, potholders, twist ties, pepper grinder, food mill, rolling pin, cookie cutter, colander, flour sifter); cleaning equipment (broom, dustpan, vacuum cleaner, buckets, sponges, rubber gloves, clothes pins); medical equipment (stethoscope, blood pressure cuff, slings, canes, crutches, syringes, gauze pads, adhesive tape); dress-up clothes with various types of fasteners (zipper, buttons, Velcro, snaps); and other pretend-play equipment and tools (keys, telephones, toaster, stove, desk lamp, mailbox)

Block area tools — Wood blocks, cardboard blocks, Styrofoam blocks, woodchips and shims, ruler, yardstick, measuring tape, cartons and boxes, large elastic bands, foam padding

Woodworking area tools — Screwdrivers, hammers, pliers, clamps, hand drills, levels, dowels, fasteners and other hardware (nails, screws, nuts, bolts, washers)

Toy area tools — Beads and string, corks, pegs and pegboards, game spinners, rubberbands

Sand and water table tools — Pail and shovel, buckets, plastic containers, muffin tins, mesh strainer, turkey baster, bubble wands, funnel

Outdoor play area tools — Sand and water table equipment (see above) plus rope, wagon, saw horses, dump truck, wheelbarrow, gardening tools (spade, hoe, rake, gardening gloves, kneeling pads, watering can, hose), hand pump (to inflate balls)

To expand the collection of tools in your setting, bring tools and gadgets from home. Encourage families to contribute old and unused items, such as wind-up clocks, cell phones (with batteries removed), and carpentry tools. Visit garage sales and flea markets to find familiar and unusual tools and gadgets. Encourage children to think about the purpose and uses of these various tools.

Help children consider how and why to use tools in various ways

With science learning in mind, help children as they explore different tools, reflect on how they work, and think about the ways they can be used to further their play ideas. In addition to supporting children's spontaneous discoveries about tools during work time and outside time, plan activities where they can use tools at other times of the day. Also look for opportunities for children to make and share discoveries about tools as they engage with other content areas.

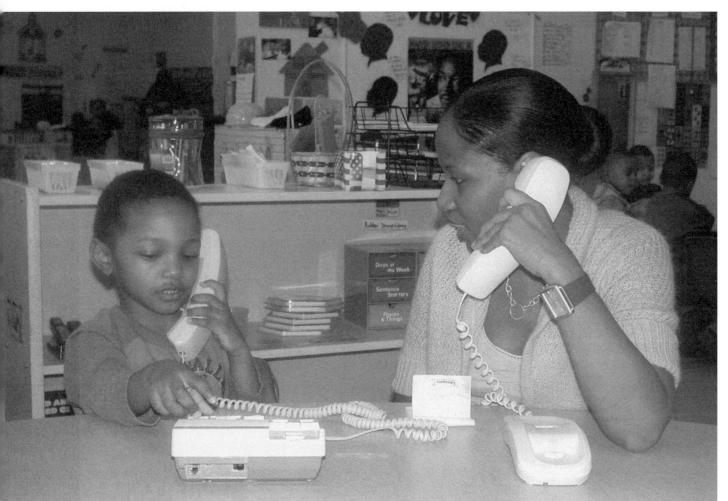

As resources for all types of scientific discovery, tools should be available in all areas of the classroom. With science learning in mind, adults can support children as they explore different tools, reflect on how they work, and think about the ways they can be used to further their play ideas.

Examples of planned activities in which children can use and think about tools include

Small-group time. Plan small-group times where children explore related tools and use them with various materials (different types of hammers, nails and golf tees, and surfaces to pound them into such as wood, clay, and Styrofoam; different tools for making imprints in clay and play dough such as hardware, kitchen utensils, toy wheels with tire treads).

Large-group time. Plan large-group times in which children use various materials as sound-making tools or instruments (banging or rubbing together foil pie tins to create different sounds; using a feather duster as a "baton"). Encourage them to think of ways to use familiar objects as "tools" to assist their movement activities (using a timer or whistle to signal stopping and starting; shaking water bottles to the beat; landing on carpet square "lily pads").

Recall. Watch for a tool each child uses during work time, give clues related to how each tool works and how it was used, and let children guess what each tool is. As they begin to think about tools more broadly (a tool is more than just a hammer or a drill), ask them to get a tool that they used during work time, bring it to recall, and describe how they used it.

Snacktime. When children use cooking utensils and tableware to prepare and eat snacks, encourage them to think about how these tools work (using tongs to pick up crackers; spreading peanut butter with a butter knife or spoon; using slotted and plain spoons to scoop fruit from a juicy bowl of fruit salad).

Field trips. Visit hands-on (children's) science museums with displays of interesting gadgets and encourage children to describe them and speculate on their uses. Visit other places with tools, such as hardware and home improvement stores, craft supply stores, kitchen stores. (First make sure the class is welcome and plan appropriate hands-on experiences with your hosts.)

Examples of other content areas in which children can use and think about tools include the following:

Creative arts. Provide tools the children can use to shape clay or play dough, paintbrushes in a variety of widths they can use and compare, and other tools listed here so they can solve problems encountered in exploring the arts. Encourage children to represent the tools they use in drawings and models (the kitchen utensils used to mix and bake "cookies"), to move like tools during movement activities (turn like a screwdriver), and to create tools during pretend play (build a "camp stove" with cartons and pie tins). Using, creating, and representing tools can help children think about how tools are made and how they work.

Reading. Read books with children about scientists and their inventions. Bring in catalogs with tools for professionals and home hobbyists (for carpenters, cooks, artists, gardeners). Listen to audio versions of books and stories.

Mathematics. Encourage children to use conventional and unconventional tools (rulers, pieces of string) to measure objects. Use tools as an aid to counting (set aside one block each time a child counts off another unit). Use tools such as clipboards, paper, and markers to conduct simple data analysis (to sort and tally the number of items collected on a walk).

Conflict resolution. Provide timing devices such as sand timers for children to keep track of the number and length of turns they take when sharing a toy. Encourage them to use tools to make substitute objects, for example, to string beads when there are not enough necklaces in the jewelry box for every child who wants to wear one.

As preschoolers experiment and become more adept at using tools, they increasingly focus on the effects of their actions and consider how a tool can help them accomplish a goal or solve a problem.

Choose and mediate children's use of appropriate technology

Adults play a critical role in selecting appropriate hardware and software and helping young children use them in constructive ways. First and foremost, remember that a technological device is only one among many types of tools that children explore and use to carry out their intentions. Therefore, computers and other electronic equipment should never dominate in the learning environment. However, because children vary widely in their exposure to technology outside the program, making it available in the classroom can provide an important experience for those whose family income or other factors limit their access (Burkham & Lee, 2002).

To help children both balance and benefit from their use of computers and related technology, do the following:

Model safe and careful use of technology

Help children learn to use technology (as they do other classroom materials) in ways that will neither hurt them nor damage equipment. For example, decide when and how children can turn machines on and off, model how to handle equipment gently without throwing or banging on it, explain why water or paint could damage it, and identify which items can and cannot be safely moved to another area of the classroom. Set clear expectations and provide time and support for children to learn what those expectations are. Help them understand that using technology properly is no different than wearing a smock to paint or safety goggles to hammer. Computers, paintbrushes, and carpentry supplies are merely different types of tools that children can use to carry out their ideas.

Choose child-friendly computer hardware

Look for equipment such as touchscreens, oversized keyboards, colored keyboard keys, and a small mouse. Model basic computer skills, such as turning on the computer and screen, and moving the mouse over the mousepad. With a conventional mouse, it may help to mark which button the children should click. Keyboards are not necessary for most programs and can often be removed altogether. Encourage children who already know how to use the equipment to help their peers.

Select appropriate computer software

Choose programs that emphasize interactive and open-ended discovery and learning rather than drill and practice. Introduce the software to a few children at a time, for example, as a small-group activity. Let children explore the software and then make it available in the classroom for those who choose to use it at work time. Check websites (other than those sponsored by the manufacturers) for recommendations on appropriate software for young children.

Locate computers to facilitate social exchanges

Allow space for more than one child as well as an adult. For example, provide two chairs in front of the computer for children to sit side by side and another alongside them for an adult. If resources permit, have more than one computer in the area so children can share their ideas. Place computers where they are visible from other areas of the classroom so children can wander over and join in. Encourage children to work together at the computer. Be available to help mediate social conflicts (see KDI 15. Conflict resolution).

Encourage children to verbalize their thinking as they solve problems using technology

Problems and questions may arise with hardware ("How do I turn on the screen?"; "How do I make the sound louder?"; "Why isn't the arrow moving?") or software ("How do I figure out which piece completes the puzzle?"). Help children reflect on their solutions if the equipment is not doing what they want, or the program's feedback says an answer is wrong. Always be available when children use technology so they do not get too frustrated or discouraged. Turn "error" messages into learning opportunities.

Balance technology with other hands-on learning

Computers and other electronic devices are only one type of tool among the many you can make available in the classroom. Technology should never overwhelm or substitute for children's hands-on learning with real materials that provide sensory feedback, or opportunities to solve social problems during other types of play.

For examples of how preschoolers engage with exploring and using tools and technology, and how to scaffold their learning at different stages of development, see "Ideas for Scaffolding KDI 52. Tools and Technology" on page 98. Use these ideas, together with those discussed above, as you play and interact with the children in your preschool program throughout the day.

Ideas for Scaffolding KDI 52. Tools and Technology

Always support children at their current level and occasionally offer a gentle extension.

Earlier	Middle	Later
Children may	*Children may*	*Children may*
• Explore tools for their own sake (e.g., put many staples in a piece of paper rather than stapling things together; pound a hammer). • Explore technology for its own sake (e.g., see what happens when they push a button on the keyboard or move their hand across a touchscreen).	• Use tools to support their play (e.g., cut and tape strips of paper together to make a "dog leash"; use a flashlight to look at books inside a tent). • Use technology to support their play (e.g., sit alongside another child and take turns touching the screen while playing a computer game).	• Explain in simple ways how tools work and what they can be used for (e.g., "You push the top down and the staple comes out here"; "The magnifying glass makes the bug look bigger"). • Explain in simple ways how a piece of technology works (e.g., "If you move the mouse here and click on this, it makes the star get bigger").
To support children's current level, adults can	*To support children's current level, adults can*	*To support children's current level, adults can*
• Provide time for children to explore tools; comment on how children use tools; copy children's actions with tools (e.g., "I'm putting lots of staples in my paper too"). • Describe children's actions on the computer and their effects (e.g., "When you pushed this button it got louder, and when you pushed the other button it got softer").	• Acknowledge what children accomplish with tools (e.g., "You counted two turns with the sand timer"; "You used the blanket like a big basket and each held one end"). • Comment on how children use technology in their play (e.g., "You made a menu on the computer. I'd like to order the tomato soup, please"); sit alongside children while they work at the computer and comment on what they are doing.	• At recall time, ask children to explain how they used tools (e.g., "How did you get your picture to stay up on the bulletin board?"). • Repeat children's explanations and directions to be sure you understand them (e.g., "If I move the mouse like this, the shape flips over, right?").
To offer a gentle extension, adults can	*To offer a gentle extension, adults can*	*To offer a gentle extension, adults can*
• Encourage children to use tools to support their play or solve problems (e.g., "Let's see if there's anything in the art area you could use to attach the ribbon to the cardboard"). • Encourage children to try new actions at the computer (e.g., "I wonder what would happen if you touched each button in this row"); discuss the effects of their actions (e.g., "That button made the cursor move sideways").	• Ask children to direct you in using a tool (e.g., "Show me how you used the hand drill so I can make a hole that looks like yours"). • Encourage children to show others how to use technology (e.g., "Brian was playing with the drawing program earlier; maybe he can show you how he turned the shapes around").	• Ask children what else a tool could be used for and why (e.g., "I wonder what else you could use the stapler for. Why do you think it will work for that?"). • Make a mistake using technology, comment that it's not working, and ask the child to help you (e.g., click the wrong icon to get a program to advance to the next level and say, "It's not changing. How do I get it to go to the next screen?").

Science and Technology: A Summary

General teaching strategies that support science and technology
- Introduce children to the steps in the scientific method.
- Encourage reflection.
- Create opportunities for surprise and discrepancy.
- Encourage documentation.
- Encourage collaborative investigation and problem-solving.

Teaching strategies that support observing
- Provide a sensory-rich environment.
- Establish a safe environment for children to observe with all their senses.
- Provide the vocabulary to help children label, understand, and use their observations.

Teaching strategies that support classifying
- Encourage children to collect and sort things.
- Call attention to *same* and *different*.
- Use *no* and *not* language.

Teaching strategies that support experimenting
- Ask and answer "What if…?" and "Why?" and "How?" questions.
- Encourage children to gradually replace trial-and-error exploration with systematic experimentation.
- Provide materials and experiences for investigating how things change with time.

Teaching strategies that support predicting
- Help children reflect on the similarities between their past and present experiences.
- Encourage children to say what they think will happen.
- Encourage children to verify their predictions.

Teaching strategies that support drawing conclusions
- Provide materials and experiences that work in similar but not identical ways.
- Encourage children to reflect on the processes and outcomes they observe.

Teaching strategies that support communicating ideas
- Use scientific language as you talk with children about their actions, observations, and discoveries.
- Provide opportunities for children to symbolically represent their scientific experiences.

Teaching strategies that support learning about the natural and physical world
- Provide materials and experiences for children to gather knowledge about the natural and physical world.
- Encourage children to make connections to explain how the world looks and functions.

Teaching strategies that support learning about tools and technology
- Provide a variety of tools in all areas of the classroom.
- Help children consider how and why to use tools in various ways.
- Choose and mediate children's use of appropriate technology.

References

Bruner, J. S., Olver, R. R., & Greenfield, P. M. (1996). *Studies in cognitive growth*. New York, NY: Wiley.

Burkham, D. T., & Lee, V. E. (2002). *Inequality at the starting gate: Social background differences in achievement as children begin school*. Washington, DC: Economic Policy Institute.

Campbell, P. F. (1999). Fostering each child's understanding of mathematics. In C. Seefeldt (Ed.), *The early childhood curriculum: Current findings in theory and practice* (3rd ed., pp. 106–132). New York, NY: Teachers College Press.

Chalufour, I., & Worth, K. (2003). *Discovering nature with young children*. St. Paul, MN: Redleaf Press.

Chess, S., & Thomas, A. (1996). Temperament. In M. Lewis (Ed.), *Child and adolescent psychiatry: A comprehensive textbook* (2nd ed., pp. 170–181). Baltimore, MD: Williams & Wilkins.

Chinn, C. A., & Schaverien, L. (1996). Children's conversations and learning science and technology. *International Journal of Science Education, 18*(1), 105–116.

Church, E. L. (2003). Scientific thinking: Step-by-step. *Scholastic Early Childhood Today, 17*(6), 35–41.

Clements, D. H. (1999). The effective use of computers with young children. In J. V. Copley (Ed.), *Mathematics in the early years* (pp. 119–128). Reston, VA: National Council of Teachers of Mathematics and National Association for the Education of Young Children.

Clements, D. H. (2002). Computers in early childhood mathematics. *Contemporary Issues in Early Childhood, 3*(2), 160–181. doi:10.2304/ciec.2002.3.2.2

Copley, J. V. (2010). *The young child and mathematics* (2nd ed.). Washington, DC: National Association for the Education of Young Children and Reston, VA: National Council for Teachers of Mathematics.

DeVries, R., & Sales, C. (2011). *Ramps & pathways: A constructivist approach to physics with young children*. Washington, DC: National Association for the Education of Young Children.

Elkind, D. (1999). Educating young children in math, science, and technology. In *Dialogue on early childhood science, mathematics, and technology education*. Washington, DC: American Association for the Advancement of Science. Retrieved from http://www.project2061.org/publications/earlychild/online/context/elkind.htm

Eshach, H., & Fried, M. N. (2005). Should science be taught in early childhood? *Journal of Science Education and Technology, 14*(3), 315–336. doi:10.1007/s10956-005-7198-9

French, L. (2004). Science as the center of a coherent, integrated early childhood curriculum. *Early Childhood Research Quarterly, 19*(1), 138–149. doi:10.1016/j.ecresq.2004.01.004

Gelman, R., & Brenneman, K. (2004). Science learning pathways for young children. *Early Childhood Research Quarterly, 19*(1), 150–158. doi:10.1016/j.ecresq.2004.01.009

Gelman, S. (1999). Concept development in preschool children. In *Dialogue on early childhood science, mathematics, and technology education*. Washington, DC: American Association for the Advancement of Science. Retrieved from http://www.project2061.org/publication/earlychild/online/context/gelman.htm

Gronlund, G. (2006). *Making early learning standards come alive*. St. Paul, MN: Redleaf Press and Washington, DC: National Association for the Education of Young Children.

Haugen, K. (2010). Learning to use tools and learning through tools: Brain development and tool use. *Exchange, 32*(5), 50–52.

Haugland, S. W. (1992). Effects of computer software on preschool children's developmental gains. *Journal of Computing in Childhood Education, 3*(1), 15–30.

Hyson, M. (Ed.). (2003). *Preparing early childhood professionals: NAEYC's standards for programs*. Washington, DC: National Association for the Education of Young Children.

Katz, L., & Chard, S. C. (1996). *The contribution of documentation to the quality of early childhood education*. Retrieved from ERIC database. (ED393608 1996-04-00)

Landry, C. E., & Forman, G. E. (1999). Research on early science education. In C. Seefeldt (Ed.), *The early childhood curriculum: Current findings in theory and practice* (3rd ed., pp. 133–158). New York, NY: Teachers College Press.

Langer, J., Rivera, S., Schlesinger, M., & Wakeley, A. (2003). Early cognitive development: Ontogeny and phylogeny. In J. Valsiner & K. Connolly (Eds.), *Handbook of developmental psychology* (pp. 141–171). London, England: Sage.

Lind, K. K. (1999). Science in early childhood: Developing and acquiring fundamental concepts and skills. In *Dialogue on early childhood science, mathematics, and technology education.* Washington, DC: American Association for the Advancement of Science. Retrieved from http://www.project2061.org/publications/earlychild/online/experience/lind.htm

National Committee on Science Education Standards and Assessment (NCSESA), National Research Council. (1996). *National science education standards.* Washington, DC: National Academies Press.

Neill, P. (2008). *Real science in preschool: Here, there, and everywhere.* Ypsilanti, MI: HighScope Press.

Piaget, J. (1950). *The psychology of intelligence.* London, England: Routledge.

Piaget, J. (1951/1962). *Play, dreams, and imitation in childhood.* New York, NY: Norton.

Piaget, J. (1955). *The language and thought of the child.* New York, NY: World Publishers.

Russell, T., Harlen, W., & Watt, D. (1989). Children's ideas about evaporation. *International Journal of Science Education, 11,* 566–576.

Seefeldt, C., & Galper, A. (2002). *Active experiences for children: Science.* Upper Saddle River, NJ: Pearson Education.

Tudge, J., & Caruso, D. (1988, November). Cooperative problem-solving in the classroom: Enhancing young children's cognitive development. *Young Children, 44*(1), 46–52.

Van Scoy, I. J., & Fairchild, S. H. (1993). It's about time! Helping preschool and primary children understand time concepts. *Young Children, 48*(2), 21–24.

Vygotsky, L. (1978). *Mind and society: The development of higher psychological processes.* Cambridge, MA: Harvard University Press.

Wilson, R. (2002). Promoting the development of scientific thinking. *Early Childhood News.* Retrieved from http://www.earlychildhoodnews.com/earlychildhood/article_view.aspx?ArticleID=409

Worth, K., & Grollman, S. (2003). *Worms, shadows, and whirlpools: Science in the early childhood classroom.* Portsmouth, NH: Heinemann and Washington, DC: National Association for the Education of Young Children.